PRAISE FOR *A BRIEF HISTORY OF TIMEKEEPING*

"As Chad Orzel shows in his informative new book, while the pace of modern life seems to march briskly in step with the rhythms of various clocks, keeping accurate time has been a mainstay of history—a driving force for astronomical measurements, and eventually classical and relativistic physics. *A Brief History of Timekeeping* offers the quintessential account of all the factors that make up ways we record time—from the relatively slow progression of daily and lunar cycles to the near-instantaneous speed of atomic transitions. Orzel's fascinating chronicle of how we measure the seconds, days, and years that set the stride of our life's journey is well worth making the time to read—and that literary adventure will fly by, no doubt."

—**Paul Halpern, author of *Flashes of Creation: George Gamow, Fred Hoyle and the Great Big Bang Debate***

"Orzel gives us the grand tour of something we all take for granted. It's about time."

—**Chris Ferrie, coauthor of *Where Did the Universe Come From? And Other Cosmic Questions* and author of *Quantum Physics for Babies***

"Each day in 2019, Chad Orzel informs us, is nearly two milliseconds longer than days were in 1870. And they feel even longer. This entertaining and engrossing book takes us through our long struggle to measure time with precision. Filled with amazing devices, it's ultimately a story of the triumph of human ingenuity."

—**Sean Carroll, author of *Something Deeply Hidden: Quantum Worlds and the Emergence of Spacetime***

"A deliciously detailed journey through the astonishing ticks and tocks of timekeeping, from neolithic henges and Mayan number systems to cinnamon-filled sandglasses, tuning fork wristwatches, and even the northern lights. Equal parts mesmerizing and fascinating, Orzel's beautifully clear explanations of physics illuminate subjects from planets to quantum engineering. By the end it is clear that time may never be on our side, but keeping track of it has opened up the universe for us."

—Caleb Scharf, author of *The Ascent of Information: Books, Bits, Genes, Machines, and Life's Unending Algorithm*

"If there's one feat our species has truly mastered, it's slicing and dicing time into ever-shorter intervals, and keeping track of those ticks and tocks with mind-boggling precision. Today's best atomic clocks can track time with a precision of one part in a billion—but getting to this point, as Chad Orzel's entertaining new book shows, has been an incredible adventure. It's a history of technology, of course, but we also learn about the underlying science, from the ancient astronomers who first made sense of the motions of the sun, moon, and stars to those who unveiled relativity and quantum mechanics in the last century. If you like science, history, and fun in equal measure, *A Brief History of Timekeeping* is for you."

—Dan Falk, science journalist, author, and broadcaster

"A fascinating intersection of science, history, and theology. I never expected to lose track of time reading a book about time."

—James Breakwell, comedy writer, creator of @XplodingUnicorn on Twitter, and author of *How to Be a Man (Whatever That Means)*

A BRIEF HISTORY OF TIMEKEEPING

ALSO BY CHAD ORZEL

Breakfast with Einstein: The Exotic Physics of Everyday Objects
Eureka: Discovering Your Inner Scientist
How to Teach Relativity to Your Dog
How to Teach Quantum Physics to Your Dog

A BRIEF HISTORY OF TIMEKEEPING

The Science of Marking Time, from Stonehenge to Atomic Clocks

CHAD ORZEL

BenBella Books, Inc.
Dallas, TX

BenBella Books, Inc.
10440 N. Central Expressway
Suite 800
Dallas, TX 75231
benbellabooks.com
Send feedback to feedback@benbellabooks.com

BenBella is a federally registered trademark.

Printed in the United States of America
10 9 8 7 6 5 4 3 2 1

Library of Congress Control Number: 2021040500
ISBN 9781953295606 (trade paper)
ISBN 9781953295941 (electronic)

Editing by Laurel Leigh and Alexa Stevenson
Copyediting by Scott Calamar
Proofreading by Denise Pangia and Cape Cod Compositors, Inc.
Indexing by WordCo Indexing Services, Inc.
Text design and composition by Aaron Edmiston
Cover design by Pete Garceau
Cover image © iStock / mladn61, robas, Wylius, and Nastco
Printed by Lake Book Manufacturing

For my grandmother, Ann Ryan,
who is such a pivotal part of our family history.

Contents

A CLOCK IS A THING THAT TICKS

The picturesque campus of Union College in Schenectady, New York, where I teach, features lush, green quads surrounded by columned buildings full of classrooms, labs, and offices. Like most college campuses, it's also full of clocks. There's the tower clock in Memorial Chapel, whose chimes regularly ring out to mark the hours (and showcase the talents of the student musicians who occasionally take over to play tunes on the chimes). A large decorative clock, a gift of the class of 1997, stands in front of the Reamer Campus Center, where it serves as a landmark for meeting people. Almost every classroom sports an analog clock on the wall, for students and faculty to track the progress of class hours (too fast for the faculty, too slow for the students). And, of course, there are innumerable unseen clocks: embedded in every computer, worn as wristwatches, carried in smartphones.

Even beyond the physical presence of clocks, the tracking of time is ubiquitous on campus. Classes are scheduled down to the minute—start at 9:15 AM, end at 10:20 AM, start again at 10:30 AM in a different room—and meetings and appointments fill the day. Student work is often timed—10 minutes allotted for an in-class presentation, an hour to complete an exam—and athletic feats are recorded down to fractions of a second. On a slower scale, the academic year progresses through a stately cycle of marked dates: the opening convocation, the start and end of the academic term, the annual Steinmetz Symposium celebrating student research. Every day brings students and faculty closer to the

1

commencement ceremony that ends the year and sends a new class of graduates off into the world.

All of this tracking and measuring of time is generally regarded as too routine to be commented upon. What's too often overlooked in the process of keeping on schedule (and fretting about being overscheduled) is the depth of history and rich variety of science underlying all this activity. We tend to think of our preoccupation with time as a modern phenomenon, but in fact the tracking of time has been a major concern in essentially every era and location that we find evidence of human activity. Every historical society we know of had its own ways of tracking the passage of time, some of them dizzyingly complicated. Some of the most ancient architectural monuments we know of are calendar markers.

A Brief History of Timekeeping is a virtual journey through the science that grew up alongside centuries of human efforts to measure the passage of time. Starting with the oldest known solstice-marking monuments, we will explore the astronomy of the solar system and the features that determine the sun's path across the sky. Efforts to better understand the motion of the sun and planets led to the development of Newtonian physics, and we'll see how this enables the technology to build mechanical clocks. And we'll discuss revolutionary developments in the physics of electromagnetism and quantum mechanics that eventually led to the replacement of standards of timekeeping based on the motion of physical objects (like the pendulum of a mechanical clock) by modern atomic clocks that count the oscillations of light.

The history of timekeeping is not only a story of abstract science and technology: it also includes intriguing elements of politics and philosophy. We'll look at the social aspects of calendar systems like the intricate mix of cycles in the Mayan calendar at their peak around 500 CE, the theological considerations that led Europe to adopt the Gregorian calendar nearly a thousand years later, and the political negotiations that produced our modern system of time zones. And we'll see how practical issues involved in these processes raised philosophical questions that in turn reshaped our understanding of our place in the universe as well as the nature of space and time.

The tracking of time is, in a deep way, a signature preoccupation not just of modern society, but of human civilization in general. The process of building and refining timekeeping devices has been one of the great drivers of progress in

A SIDEBAR REGARDING BARS ON THE SIDE

As you read through this book, you will encounter occasional sections high-lighted, as this one is, by shaded bars along the sides of the page. These sections go a little deeper into the scientific principles underlying particular methods of timekeeping and have less historical content.

I've made every effort to make *all* the sections of the book approachable and engaging for as broad an audience as possible—you won't find extensive math or undefined jargon—but recognize that individual readers may find particular topics more or less congenial. While I hope you'll read every word on every page, the shading is a convenient way to mark sections that are slightly more technical. Those who are primarily interested in the scientific content may want to seek out the highlighted sections; those who are primarily inter-ested in the historical aspects may want to pass over them a bit more lightly. (There are also some shorter boxes, which house topics or anecdotes that fall to the side of the book's primary thrust but were too good not to share.)

science and technology for millennia: from Neolithic solstice markers through mechanical watches to ultra-precise laser frequency standards, we are and always have been a species that builds clocks.

Of course, a statement that sweeping depends on a particular definition of terms. In this case, we need a definition of "clock" that encompasses all these devices over a span of several thousand years. So, what is the common feature shared by solstice markers, mechanical watches, and laser spectroscopy? At the most basic level, a clock is a thing that ticks.

The "tick" here can be the audible physical tick we associate with a mechanical clock like the one in Union's Memorial Chapel, caused by collisions between gear teeth as a heavy pendulum swings back and forth. It can also be a more subtle physical effect, like the alternating voltage that provides the time signal for the electronic wall clocks in our classrooms. It can be exceedingly fast, like the nine-billion-times-a-second oscillations of the microwaves used in the atomic clocks that provide the time signals transmitted to smartphones via the internet, or ponderously slow like the changing position of the rising sun on the horizon.

In every one of these clocks, though, there is a tick: a regular, repeated action that can be counted to mark the passage of time. When the microwave field in an atomic clock completes an oscillation from up to down and back, we know that 1/9,192,631,770th of a second has passed. When the shadow cast by a building on campus sweeps across the quad and then returns to pointing due north, we know that one full day has passed. When the rising sun completes its seasonal cycle and returns to its northernmost position on the horizon, we know that one full year has passed. Each of these events is a metaphorical tick, and we quantify the passage of time by counting the number of these ticks that happen between two events of interest.

It follows that one of the core themes in the history of timekeeping is the process of identifying and understanding these ticks. To make a high-quality clock, we must first identify a process that might serve as a tick, and then understand all the factors that change its rate and how those factors affect the ultimate accuracy of that clock. The drive to understand the processes by which various kinds of clocks tick has led to revolutionary discoveries in fundamental science, particularly in physics and astrophysics (as we'll see in Chapters 6 and 8).

In order to time events by counting ticks, a practical clock must also have some sort of readout, a means by which humans can keep track of the count. While the audible ticks of a mechanical clock can, in principle, be counted directly, many of the physical systems used as clocks work at speeds that are impractical for human observers; the quartz crystal in my wristwatch oscillates much too rapidly to be counted directly, so its motion must be recorded electronically and slowed down to move the second hand one mark for every 32,768 vibrations.

Many of the signature features we associate with practical clocks are these readout devices: the hands and numbers on a watch, or the gnomon and markings of a sundial tracking the slow motion of a shadow through the day. The history of timekeeping is thus also a history of innovation in readout technologies, finding clever ways to display the passage of time in human-readable ways that minimize sources of error. One of the most dramatic examples is the competition between methods for finding longitude at sea (described in Chapters 9 and 10), a contest decided in part on the ability of ordinary sailors to use them.

Another problem plaguing the makers of clocks through the centuries is that many excellent standards are impractical to transport. A solstice marker is fundamentally bound to a particular point in space, recording the position of the sun as seen from a particular location; if you shift the marker to another position, its references no longer apply. At the opposite technological extreme, a precision atomic clock requires a significant infrastructure for the preparation and interrogation of atoms, keeping the very best atomic clocks confined to laboratories.

Many of the practical clocks we surround ourselves with are thus actually *models* of clocks, a way of keeping track of time that approximates the time recorded by a better standard tick that is inconveniently immobile. The clock on the wall of my classroom displays only an approximate time based on the oscillation of the electric current supplied from the power grid; the official time is determined by an assortment of atomic clocks in national standards laboratories scattered around the globe and overseen by an international governing body (we'll discuss the origin of global time zones and the determination of the official time in Chapter 11).

Because most practical clocks are models of time, a third core theme in the history of timekeeping is the idea of *comparing* clocks to one another, and using that comparison to refine our models of time. The fundamental process is universal, employed by every civilization that has modeled time: you synchronize your model clock with some standard, allow it to run freely for many ticks, then check it against the standard again and see if they're still in synch. It's the same process you might use to check the performance of your watch: you set it to the correct official time (say, using the US time web page provided by the National Institute of Standards and Technology [NIST]), wear it around for a while, and then some days or weeks later check it against the official time again, and adjust as needed. Even though the quartz crystal in your watch is not connected in any direct way to the cesium atoms in atomic clocks at NIST, this process ensures that you carry with you a very good model of the actual official time.

Through the long history of timekeeping, this process has been used to produce dramatic results on a variety of timescales. As we'll see in Chapter 12, the process of synchronizing two clocks at different locations through the exchange of telegraphic signals was central to the development of the theory of relativity, with momentous consequences for our understanding of time and space. At the other extreme of duration (as we'll see in Chapter 3), the Gregorian calendar reform grew out of this synchronize–free-run–check process applied over the course of centuries.

The Gregorian calendar is also an important illustration of a fourth theme: the way the sheer depth of the history of timekeeping enables incredible precision, even with crude tools. Thanks to calendar records spanning thousands of years, from Roman times through the medieval period and into the Renaissance, astronomers could detect a difference between their model of time and the cycle of the seasons that amounts to just 11 minutes *per year*. Following the "synchronize" phase of calendar reform in 8 CE, the Julian calendar ran freely for more than 1,500 years before its final "check," which led to the introduction of our modern calendar by Pope Gregory XIII in 1582. The Gregorian calendar will stay in synch with the seasons for another 3,000 years, despite being implemented two decades before the invention of the telescope. You don't need atomic clocks with nanosecond precision to make highly accurate measurements of time, provided you can wait long enough.

Pulling all these elements together, we can see the history of timekeeping as a progression through a long series of standard ticks and tools to model them. As our knowledge of science has advanced, we have found new ways to measure more subtle effects, allowing a move to better standards through a kind of bootstrapping process: a new candidate tick is first verified against the best version of the current standard, and then as the processes behind the new standard become better understood, it eventually surpasses the precision of the current standard, and then supplants it.

This is nicely encapsulated in the changes in the official definition of time through the twentieth century: when what became the modern International System of Units (SI) was introduced in the 1870s, one second was defined as 1/86,400th of a solar day. This was an adequate definition for clocks at that time, but the rotation rate of the earth changes over time: a solar day in 2019 is nearly two milliseconds longer than a day in 1870. As state-of-the-art pendulum clocks and quartz crystal oscillators were developed in the early 1900s, their precision advanced to the point where the intrinsic uncertainty of these model clocks was smaller than the ambiguity introduced by the definition of a "day." Removing this ambiguity required a change in the definition of time.

In 1960, the definition of a second was officially changed to "the fraction 1/31,556,925.9747 of the tropical year for 1900 January 0 at 12 hours ephemeris time," tying the second to the well-understood motion of the earth about the sun. Of course, it's impossible to keep a copy of the tropical year 1900 around for reference, so physicists looked for a more convenient and universal standard. They eventually found one thanks to the quantum nature of atoms and light: the frequencies of light absorbed and emitted by atoms are uniquely determined by the laws of quantum physics, and every atom of a given element is identical to every other, making atoms a convenient and somewhat portable (more so than the year 1900, at least) standard for timekeeping.

With the completion of quantum theory and the development of high-precision microwave spectroscopy (which we will discuss in Chapter 13), it became possible to decouple the definition of time from the slowly changing motion of the earth. In 1967, the definition of the second was changed to "the duration of 9,192,631,770 periods of the radiation corresponding to the transition between the two hyperfine levels of the ground state of the cesium-133 atom." This has remained the official definition, the tick that we count to track what time it is, and the best atomic clocks today would need to run continuously for a billion years or more before they would gain or lose a single second.

The cesium standard is not necessarily the *final* definition, though, as physicists and engineers have continued to experiment with newer and better types of clocks. There are experimental clocks under development in standards laboratories tens or hundreds of times more accurate than the best possible cesium standards. A clear favorite has yet to emerge, but not too many years in the future, we may see yet another redefinition of time, moving away from cesium to an entirely different element (we'll discuss these next-generation atomic clocks in Chapter 16).

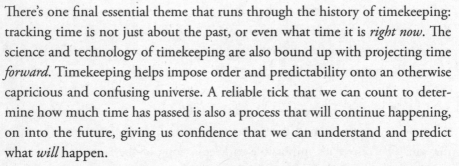

There's one final essential theme that runs through the history of timekeeping: tracking time is not just about the past, or even what time it is *right now*. The science and technology of timekeeping are also bound up with projecting time *forward*. Timekeeping helps impose order and predictability onto an otherwise capricious and confusing universe. A reliable tick that we can count to determine how much time has passed is also a process that will continue happening, on into the future, giving us confidence that we can understand and predict what *will* happen.

This forecasting element shows up throughout the history of timekeeping, all around the world. We see the evidence of Neolithic cultures setting up monuments to mark the winter solstice and provide a reminder of the spring to come. Jews and early Christians adjusted their calendar systems in order to predict the dates of Passover and Easter years into the future. Mayan astronomers tracked the motion of Venus to predict auspicious times for war. European mathematicians developed elaborate models of where the moon *will be* to help sailors navigate their way through globe-spanning empires. There's even an element of this forward projection in modern efforts to push the frontiers of atomic clocks: we don't need that level of precision right now, but it could provide the key to exotic discoveries in the future.

Through all this history, from ancient agrarian societies looking for the best time to plant crops, to astrologers trying to read the future in the stars, to astronomers predicting transits of celestial objects, timekeeping has always been as much about the future as the past. That predictive element shows up again and again, from the most ancient builders of stone monuments all the way to modern office workers entering future vacations into digital calendar apps.

In modern society, we expend an enormous amount of effort on the tracking of time, supported by a vast and largely overlooked foundation of science and technology. The development of that foundation is a story that spans several thousand years and incorporates an enormous amount of scientific knowledge. That's a history that's worth taking a little time to explore.

Chapter 1

SUNRISE

In the late 1600s, the "Glorious Revolution" deposed the Catholic king of England and Scotland, James Stuart, replacing him with his daughter, Mary, and her husband, William of Orange. This triggered a number of uprisings in the British Isles, including the "Williamite War" in Ireland. After the uprisings were put down, large tracts of land were confiscated from James's Catholic supporters and redistributed among William's Protestant backers. One of these parcels, an estate near the port town of Drogheda, passed to a man named Charles Campbell, who set about making improvements to his new property. To this end, in 1699 he ordered a crew of laborers to begin quarrying stones from an overgrown rocky mound on a hilltop.

This mound was one of many dotting the landscape in the area known as the Brú na Bóinne ("Palace of the Boyne"), which was strongly associated with the gods and heroes of Ireland's pagan past. Not far into their quarrying work, Campbell's crew uncovered a "broad flat Stone, rudely Carved, and placed edgewise at the Bottom of the Mount," blocking the entrance to an artificial passage fashioned from enormous stones standing on end. This discovery brought the quarrying to a halt and attracted the attention of Edward Lhwyd, the keeper of the Ashmolean Museum at Oxford, who happened to be traveling through the area. Lhwyd described his visit to the "cave" in a letter to the Royal Society, which is the first published description of the passage tomb now known as Newgrange (and the source of the quoted description above).

Over the next couple of centuries, antiquarians and treasure hunters made occasional visits to Newgrange in the haphazard manner of the times, thoroughly muddling the archaeological record and producing several confused and mildly contradictory accounts of the tomb and its contents. The large-scale structure of the mound was largely undisturbed, though, until a massive excavation was undertaken in the 1960s by Irish archaeologist Professor Michael O'Kelly, who carefully removed the tons of earth and rock covering the passage and central vault, and in the process made a remarkable discovery about the purpose of the structure.

While excavating the entrance, O'Kelly discovered the "roof-box," two parallel slabs of rock placed horizontally just above the entrance to the passage. The gap between these slabs was closed by blocks of white quartz; marks on the stones suggested these had been removed and replaced repeatedly. A local tradition related to the O'Kelly team during the excavation held that "at some unspecified time," the rising sun used to light up a carved stone at the very back of the tomb. On the basis of this legend and the site orientation surveys done during the course of the excavation, members of the team entered the central chamber in the predawn hours of December 21, 1969, and recorded a remarkable phenomenon:

> At exactly 8.54 hours GMT the top edge of the ball of the sun appeared above the local horizon and at 8.58 hours, the first pencil of direct sunlight shone through the roof-box and along the passage to reach across the tomb chamber floor as far as the front edge of the basin stone in the end recess. As the thin line of light widened to a 17-cm band and swung across the chamber floor, the tomb was dramatically illuminated and various details of the side and end recesses could be clearly seen in the light reflected from the floor. At 9.09 hours, the 17-cm band of light began to narrow again and at exactly 9.15 hours, the direct beam was cut off from the tomb. For 17 minutes, therefore, at sunrise on the shortest day of the year, direct sunlight can enter Newgrange, not through the doorway, but through the specially contrived slit that lies under the roof-box at the outer end of the passage roof.*

---

* From O'Kelly's book, *Newgrange: Archaeology, Art and Legend.* Thames and Hudson, 1982.

TO EVERYTHING THERE IS A SEASON, TURN, TURN, TURN

We don't necessarily think about a solstice marker like Newgrange as a "clock," because of its monumental scale and the fact that the associated "tick" is so slow, occurring just once per year. This is not a device that will help you determine how much longer you have to endure a tedious lecture or wait for the arrival of a dinner companion. It unquestionably fits the broad definition spelled out in the Introduction, though: the appearance of sunlight in the central chamber provides a clear readout for a physical cycle that repeats at regular intervals, and can be counted to track the passage of time.

The physical process that provides the tick for a solstice marker like Newgrange is the motion of the sun through the seasons of the year. For people who live in the middle latitudes, like the builders of Newgrange or residents of the Northeastern United States like myself, the year is a march through winter, spring, summer, and fall, returning to winter again. These changing seasons are associated with changes in the length of the day: in the winter, the sun rises late and sets early, while in the summer, it does the opposite. As the caregiver for an active dog who needs multiple daily walks, I'm acutely aware of this; during the winter months, from November through March, both our post-breakfast and post-dinner walks take place in darkness.

During the short days of winter, the sun shines for less of the day and thus has less time to heat the surface of the earth, leading to colder daily temperatures. In the summer, longer days mean more heating and higher daily temperatures.* The seasonal cycle of weather is accompanied by seasonal changes in plants and animals: as the days get longer and the weather warmer, plants begin to grow and animals become more active, while as the days get shorter and the weather colder, trees lose their leaves and animals hunker down for the winter. These seasonal cycles are of critical importance to human civilization even today with global transportation networks ensuring access to fresh food year-round; they would've been a matter of life and death for the agrarian Neolithic society that built Newgrange.

---

* There's a bit of a time lag in the cycle—the hottest days of summer are generally a few weeks after the longest day of the year, and the coldest (and snowiest) days of winter come after the shortest day has passed. This happens because the earth is huge and it takes some time for heat to accumulate or dissipate.

This very obvious seasonal change in the length of the day is accompanied by a somewhat less obvious change in the positions of sunrise and sunset. While we learn as children that the sun rises in the east and sets in the west, this is strictly true for only two days of the year: only on the equinoxes in late March and late September does the sun rise *due* east and set *due* west.* During the short days of winter, the sun rises south of east and sets south of west, while the long days of summer see the sun rise north of east and set north of west.†

This pattern in the positions of the rising and setting sun would be unmistakable to an ancient agrarian society, but the mechanism behind it was surely a mystery to them. In modern times, we know that the sun is not actually moving but only appears to move because the earth is approximately spherical and rotates on an axis that is tilted relative to its orbit around the sun. The daily east-to-west motion of the sun across the sky is due to the rotation, while the yearly motion of the rising and setting sun along the horizon is due to a combination of the axial tilt and the orbital motion.

Daily Rotation

The daily motion of the sun across the sky is the simpler of the two motions to explain. Seen from a vantage point floating above the north pole, the earth rotates counterclockwise, so a day looks like the image on the next page.

From the perspective of someone standing on the surface of the earth facing south, the sun moves from left to right across the sky, setting in the west, whereupon the stars appear and follow a similar left-to-right trajectory through the course of the night. In the morning, the sun rises in the east, and the familiar cycle repeats.

The roundness of the earth adds a slight subtlety to this picture, one that many people find surprising. When I teach a class about timekeeping at Union

---

* The March equinox is often called the "vernal equinox" and the September one the "autumnal equinox," reflecting the seasons in which they occur in Europe. I'll stick with the month names here, though, out of respect for the feelings of potential southern hemisphere readers.

† In the mid-latitudes of the northern hemisphere, anyway, where both Newgrange and my home are located. In the mid-latitudes of the southern hemisphere, the relationship between directions and seasons is reversed: the sun rises north of east and sets north of west in the winter. In the tropics, the pattern is more complicated, as we'll see in Chapter 4.

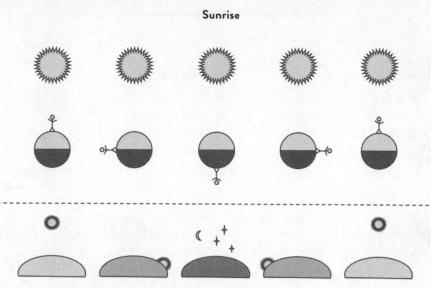

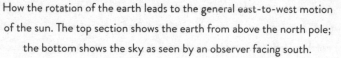

How the rotation of the earth leads to the general east-to-west motion
of the sun. The top section shows the earth from above the north pole;
the bottom shows the sky as seen by an observer facing south.

College, I have the students take a brief online quiz before the first class: one of
the questions asks where one would expect to find the sun at noon in Schenectady. A substantial fraction of the class will answer "directly overhead," in spite
of the fact that this is *never* the case. At the latitude of Schenectady, the sun at
noon is always located in the southern half of the sky.

The reason for this has to do with the spherical shape of the earth. As you
move to different positions on the surface of the earth, the direction of "directly
overhead" changes. "Up" and "down" are determined by the local force of gravity, which by definition pulls toward the center of the earth, so it *feels* the same
no matter where we are, but if you drew a line out from the earth in the "up"
direction, it would arrive at different points out in the stars depending on where
you were on the sphere. As generations of delighted children have noted, "up"
for somebody in Australia is pointing out into space in almost exactly the opposite direction of somebody in New York.*

This change in "up" means that while a person standing squarely on the
equator during one of the equinoxes will see the sun directly overhead at

---

* To be in *exactly* the opposite direction, the other person would need to be on a boat in the
  ocean several hundred miles southwest of Australia, but that's close enough for grade school.

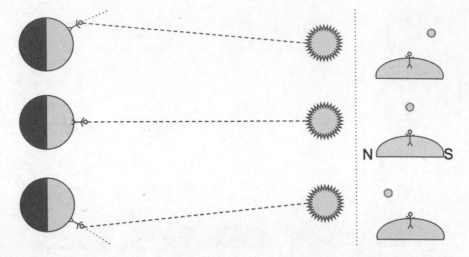

The highest point reached by the sun depends on latitude. The sun at noon is directly
overhead to an observer on the equator but displaced to the south for an observer
at northern latitudes, and to the north for an observer in southern latitudes.

midday, an observer at any other latitude will see the sun to one side or the
other. Anyone north of the equator will need to look south to find the midday
sun, while anyone at a southern latitude will find it to the north. This also means
that the sun follows distinctly different paths in the two hemispheres. At northern latitudes, the rising sun moves up and to the right (toward the south), while
at southern latitudes it moves up and to the left (north), and a similar pattern
holds at sunset.* Only on the equator does the sun go straight up from the eastern horizon, passing overhead before setting straight down in the west.

The displacement of the midday sun from directly overhead means there
will always be a discernible shadow at noon, which provides the first piece of a
process for locating yourself on the surface of the earth. The length of the noon
shadows increases as you move away from the equator, so if you measure the
length of the shadow cast by an object whose height you know, you can do a

---

* Knowing a little bit about the astronomy of the earth's rotation lets you detect a classic lazy-
film-crew trick: simulating a sunrise over the Atlantic Ocean by filming the sun setting over
the Pacific, and playing it backward. A reversed Hollywood sunset will seem to show the sun
moving up and to the left, which is the wrong direction for a New York sunrise.

little math to determine how far away you are from the place where the sun is directly overhead.*

Tilts and Seasons

The shadow-measuring trick determines the angle between "straight up" at your location and "straight up" at the spot where the sun is directly overhead, which is *almost* the latitude of your location. It's only *exactly* your longitude on two days a year, the equinoxes in March and September. Just as the direction of "straight up" changes when you move to a different place on the earth, the direction from the earth to the sun changes as the earth moves around its orbit. Because the earth's axis is tilted, this makes the point where the sun is directly overhead move around over the course of the year, and this gives rise to the cycle of the seasons.

The axis about which the earth rotates (an imaginary line connecting the north and south poles) is tilted by an angle of about 23.5 degrees relative to the axis of the earth's orbit around the sun. The direction of the axis remains constant throughout the year—a line drawn from the north pole out into space will always come close to the "North Star," Polaris, in the constellation Ursa Minor—in much the same way that the axis of a spinning gyroscope remains constant as it's moved from place to place.[†] As the earth moves around the sun, the north pole goes from being inclined slightly toward the sun to directly away from it, and everything in between.

If we start at the June solstice, which happens to be close to the point where the earth is farthest from the sun in its orbit, the north pole is inclined directly toward the sun. As the earth moves in its orbit, the angle between the direction of the axis and the direction to the sun increases, reaching 90 degrees in late September. In late December, the north pole is pointed directly away

---

* Unless you're at the equator, in which case the *absence* of a shadow will give you your location without doing any math.
† There is, in fact, a very slow drift in the direction of the earth's axis (the technical term is "axial precession"), which wobbles around a circular path over a period of 26,000 years. This change is significant on the timescale of human civilization: in 3000 BCE when Newgrange was new, the "North Star" would've been Thuban, a star in the constellation of Draco. Within any single human lifetime, though, the change is insignificant, so we can safely ignore it for our purpose.

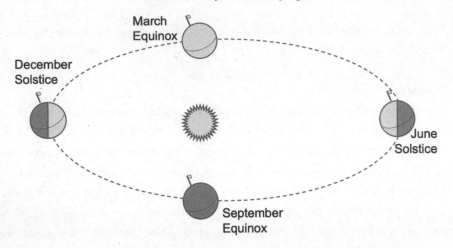

The orientation of Earth at key points in the orbit.

from the sun (and the earth is at its closest to the sun), an angle of 180 degrees, and after that the angle decreases, reaching 90 degrees in March, and coming back into exact alignment with the earth-to-sun line the following June.

There are four special points in this orbital cycle, two of which are obvious: the June and December solstices when the north pole is inclined exactly toward or exactly away from the sun. The other two are those March and September dates when the angle between the axis and the earth-to-sun line is 90 degrees. On these right-angle days, the motion of the sun is at its very simplest: it rises due east, sets due west, and the midday sun is directly overhead at the equator. No matter where you are on the earth on these days, exactly half of the rotation period is spent in light and half in darkness, which gives these dates their name: equinoxes, from the Latin for "equal night."

On any other day of the year, the combination of axial tilt and orbital motion means that one pole is inclined more toward the sun than the other, and areas closer to that pole will get more light. This accounts for the long days and short nights of summer and essentially the reverse in winter. For points very near the poles and dates close to the solstices, this can go to extremes: exactly at the poles, the sun will rise at one equinox and set at the other, remaining either above or below the horizon for six months at a time. At any point between one of the poles and the Arctic or Antarctic (at around 66.5 degrees north and south latitude), there will be at least one 24-hour period in which the sun never sets, and another when it never rises.

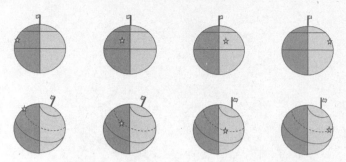

The rotation of the earth for one of the equinoxes (top) and one of the solstices (bottom).

In terms of the daily motion of the sun across the sky, the effect of this axial tilt is to change the latitude of the point on the earth where the sun is directly overhead at midday. Between March and September, this point is at a latitude north of the equator, and between September and March, the fall and winter months for the northern hemisphere, this point is at a latitude south of the equator. This is what defines "the tropics" in geographical terms: the Tropic of Cancer is 23.5 degrees north latitude, where the sun is directly overhead at noon on the solstice in late June, and the Tropic of Capricorn is 23.5 degrees south latitude, where the sun is directly overhead at noon on the solstice in late December.

For people living between the Tropic of Cancer and the Arctic Circle, like the Neolithic builders of Newgrange or modern Americans, the sun always both rises and sets every day, and it's always in the southern half of the sky. The maximum height it reaches changes depending on the season, though, moving higher in the summer and lower in the winter. In terms of the sun's path across the sky, summer is like moving closer to the equator for a few months, while winter is like moving closer to the pole. This also complicates the trick of using shadows to find your latitude; you can easily measure a *difference* in latitude from the length of a midday shadow, but converting to an exact position requires knowing where on the earth the sun is directly overhead on that date.

These changes in the length of the day and the height of the sun are accompanied by a change in the rising and setting positions of the sun along the horizon. We can understand this if we imagine looking at the earth from the sun and tracing the path followed by a particular point on the earth's surface (Schenectady, New York, say) as it rotates through the lit hemisphere. The path we see from out in space must mirror that of the sun across the sky as seen by

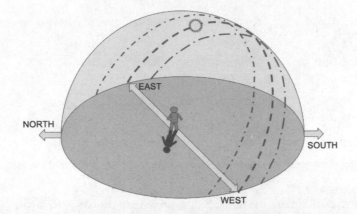

The path of the sun across the sky in the northern hemisphere on different
dates, showing that it is always in the southern half of the sky, even in
summer when it rises north of east and sets north of west.

a person on the ground at that spot on the earth. On the equinox, a point in the
northern hemisphere will move across the lit face of the earth along a straight
line. On the June solstice, however, the path for that same point is a curve that
starts to the north at sunrise, moves south during the day, then returns to the
north at sunset. A person at that point looking at the sun would see the rising
and setting of the sun taking place north of due east and due west. On the
December solstice, the curve is reversed, and the rising and setting sun have
moved south.

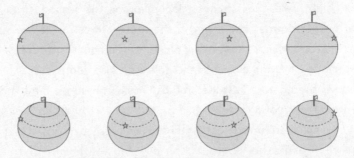

Top: Four snapshots of the lit face of the earth seen from the sun on one of the equinoxes,
with a northern hemisphere location moving straight across as the earth rotates. Bottom:
The same four snapshots on the June solstice, with the same spot following a curved path.

This motion of the sun is what the ingenious builders of Newgrange set out to capture with tons of stone. On the winter solstice, the rising sun is as far south as it will ever be through the course of the year, and the passageway from the roof-box to the central chamber is carefully aligned with this position. The Newgrange monument is built to mark the shortest day of the year, but just as importantly, it serves as a predictor of the summer solstice to come. Marking the shortest and coldest day of the year reminds us that longer and warmer days are ahead, a much-needed sign of hope for an agrarian society at the leanest time of year.

The central passage of the Newgrange tomb rises slightly up the slope of the original hill, meaning that the floor of the central chamber is barely below the elevation of the roof-box. If you drew a direct line from the floor through the roof-box, it would extend out to the southeastern horizon at a point quite close to the southernmost position of the rising sun. On the morning of the winter solstice (give or take a couple of days), a narrow beam of sunlight extends through the roof-box into the central chamber, providing the only natural illumination it will see all year.

Even after the O'Kelly excavation (and subsequent restoration of the structure),* many mysteries remain about the builders of Newgrange and what rituals were conducted at the site. There is one thing we know for sure: the Newgrange passage tomb is a clock, and one that continues to function perfectly more than 5,000 years after its construction.

STONEHENGE, ETC.

Newgrange is an exceptionally good and thoroughly investigated example of a passage tomb, but there are numerous similar constructions scattered throughout the British Isles, with different alignments. Some of these, like Dowth near Newgrange and Maeshowe in the Orkney Islands off the coast of Scotland, are aligned to the winter solstice sunrise. Others, like the Mound of the Hostages on the hill of Tara and the passage tomb at Knowth, both in Ireland, are aligned east-west to coincide with the equinoxes, and Bryn Celli Ddu in Wales has a passage tomb aligned with sunrise on the *summer* solstice. Marking the cycle of the sun by means of passage tombs was clearly a major activity in Neolithic Britain.

Several hundred miles to the southeast, and several hundred years after the completion of Newgrange, a different group of people embarked on an enormous construction project of their own, on Salisbury Plain in what's now

---

* The restored passage tomb is a popular and impressive tourist site where you can see a re-creation of the solstice sunrise using light bulbs positioned near the roof-box. You can also enter a lottery to be one of the lucky few invited to the central chamber at sunrise around the solstice (though they are careful to note, given Ireland's climate, that there are no refunds in the event of bad weather clouding out the sunrise).

Wiltshire, England. The resulting monument is one of the most instantly recognizable ancient sites on Earth, the collection of giant standing stones known as Stonehenge.

There is evidence of activity at the Stonehenge site dating back to around 3000 BCE, when somebody dug a circular ditch and piled up the removed earth to make a bank on the inner side.* The monument we see today came together over the next 1,500 years, in a series of phases of construction and reconstruction; all of these versions featured a northeast-southwest orientation, with an opening in the ditch and back leading to an "avenue" marked out by parallel ditches extending north and east from the site about half a kilometer, then turning sharply toward the nearby Avon River. The most strikingly ambitious phase of the construction, around 2500 BCE, saw the erection of five enormous "trilithons" (two uprights with a lintel across the top, making a shape like the

Stonehenge layout. Image by Joseph Lertola.

---

* Ironically, the arrangement of these mean that Stonehenge is not, strictly speaking, a true "henge" as the term is used in archaeology. A henge is defined by a ditch and bank, with the ditch inside the bank.

Greek letter pi) in a horseshoe pattern, with the open end facing northeast. These were then enclosed by a ring of smaller (but still massive) sarsen stones. The sarsen ring sits inside a rectangle defined by four "Station Stones," and a single "Heel Stone" sits outside the entrance, in the avenue bounded by the ditch-and-bank lines that extend to the river.

The northeast-facing arrangement of the trilithons at Stonehenge has prompted speculation about possible astronomical connections since the early 1700s, when the antiquarian William Stukeley noted that the trilithon horse-shoe and avenue faced toward the sunrise on the summer solstice. The open structure and large number of standing stones have allowed for some rather grandiose speculations as to the number of alignments in Stonehenge—you can draw on paper or imagine lines between any two stones and extend them to the horizon to correlate with all manner of celestial objects, but that doesn't necessarily mean these were significant to or even intended by the Neolithic builders of Stonehenge. However, there is broad agreement among archaeologists regarding a few of these alignments:

- The northeast-southwest alignment of the trilithon horseshoe and avenue face sunrise on the summer solstice and sunset on the winter solstice.
- The short sides of the rectangle defined by the Station Stones have the same northeast-southwest alignment.
- The long sides of the Station Stone rectangle are oriented toward the northernmost rising point and southernmost setting point of the moon.*

It's important to note that in the immortal words of Nigel Tufnel from *This Is Spinal Tap*, no one knows just who the builders of Stonehenge were, or what they were doing. There is even ambiguity about what, exactly, the monument was meant to celebrate: modern neopagans gather at sunrise on the summer solstice to see the sun rise next to the Heel Stone, but the great trilithon would

---

* The orbit of the moon around the earth is at a slight angle from the plane of the earth's orbit around the sun, so the moon's rising and setting positions on the horizon cover a wider range than those of the sun. More about this in Chapter 9.

just as impressively frame sun*set* on the *winter* solstice, and archaeological evidence suggests there were large gatherings at the site in winter.* Stonehenge may also have been one of a pair of linked ceremonial sites, as suggested by the Stonehenge Riverside Project excavations of 2003–2009: the nearby henge at Durrington Walls enclosed a circle of wooden posts with an alignment toward the sunrise on the winter solstice, and is reached by an avenue aligned closely with sunset on the summer solstice.†

Given the lack of written records left by the Neolithic cultures of Britain, we'll never know *exactly* what Stonehenge and other archaeoastronomical sites meant to their builders. Their monumental scale and careful alignment on important points in the solar cycle, though, clearly indicate that the builders were aware of the motion of the sun through the year and built structures that could mark the turning of the year in a dramatic and visible way.

The use of solar effects in monumental architecture is not unique to the prehistoric cultures of Britain, of course. Similar markers are found in a huge range of cultures, all around the world, and many have become famous seasonal tourist attractions. The Temple of Kukulcán at Chichen Itza in Yucatán, Mexico, built by the Maya around 1000 CE, draws crowds around the equinoxes to see an effect where the shadows cast by the corner of the stepped pyramid resemble a giant snake crawling down the balustrade. Fajada Butte in New Mexico is famed for the "Sun Dagger," where sunlight shining through the gaps between parallel slabs of rock marks the solstices by passing through key points on a spiral traced on the wall; this site was the work of the Chacoan culture between 1000–1300 CE.‡ Much later, and on the opposite side of the world, the seven-tiered gopuram gate at the Sree Padmanabhaswamy Temple in Kerala, India, constructed in the 1700s, is aligned so the setting sun on the

---

* Animal bones found in Neolithic rubbish dumps nearby allow archaeologists to estimate the age of the animals when they were slaughtered for food, and these suggest they were killed in winter.

† These are at a slight angle to one another because a sloping ridge to the west of Durrington Walls causes the setting sun to vanish at a slightly different point than it would were the horizon flat.

‡ Sadly, the volume of tourist traffic to the site caused erosion that led to the shifting of one of the slabs in 1989; the site is now closed to the public.

equinoxes passes through pairs of windows on the inner and outer walls of five of the levels, illuminating each in turn for a few minutes.

There is even an artificial "holiday" celebrating an accidental solar alignment created by a much more recent construction. The astronomer and science celebrity Neil deGrasse Tyson is credited with popularizing "Manhattanhenge," the phenomenon where, twice a year, the rising and setting sun aligns nearly perfectly with the grid pattern of New York City streets.* Seen from near the center of the city, this produces a very striking effect of the sun passing between skyscrapers. Other cities with regular grids of streets have their own "henge" dates, documented by enthusiastic amateur astronomers. The particular dates of these events are entirely accidental, but also entirely predictable by anyone fascinated with the seasonal motion of the sun along the horizon, which we humans have been tracking for millennia.

SHADOWS AND SPHERES

Our modern understanding of the earth as "round"—more accurately, as a spherical planet orbiting the sun and rotating about a tilted axis—necessarily underlies the above discussion of the sun's apparent motion through the year. We don't have any way of knowing how the builders of Newgrange 5,000 years ago understood and explained this celestial motion, only that they had an excellent grasp of this pattern as an empirical fact.

The notion of the earth as a sphere with a tilted axis has a surprisingly deep history of its own. The idea of a spherical Earth was accepted in classical times, and the first good measurement of its size dates back over 2,200 years, to the Greek scholar Eratosthenes.

Eratosthenes was born in Libya around 276 BCE and educated in Athens before settling in Egypt as one of the keepers of the Library of Alexandria around 240 BCE. Sometime after his arrival, he learned of a famous well in the city of Syene (modern-day Aswan) in southern Egypt, where the midday sun on the summer solstice perfectly illuminated the bottom, casting no shadows.

---

* The sunset Manhattanhenge dates fall in late May and mid-July; the sunrise dates are in early December and early January.

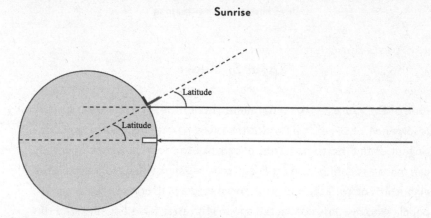

Eratosthenes's measurement principle: measuring the length of a shadow
at noon in Alexandria allowed him to calculate the angle between the
sun there and in Syene where the sun was directly overhead.

This suggests that the sun is exactly overhead in Syene on that date—in modern terms, it would be exactly on the Tropic of Cancer. Eratosthenes knew that in his home in Alexandria, the noon sun still cast a shadow on the solstice, and measuring the length of the shadow offered a way to determine the circumference of the earth.

Eratosthenes measured the minimum length of a shadow on the solstice in Alexandria, from which it's easy work for a geometer to determine the difference in latitude between Syene and Alexandria. He found that angle to be 1/50th of a full circle (7.2 degrees in modern notation). The distance from Alexandria to Syene was measured as 5,000 "stadia," in the units of the day, so Eratosthenes concluded that the full circumference of the earth must be 50 times that distance, or around 250,000 stadia. The stadia is not a terribly well defined unit by modern standards, so there is some debate about the exact result of Eratosthenes's measurement. Choosing the *worst* of the plausible values converts to a distance of about 46,000 km, which is just 15 percent larger than the currently accepted value for the circumference of the earth. That's a pretty impressive result, particularly given that Eratosthenes probably made this calculation without ever leaving his home in Alexandria!

Eppur Si Muove

One of the more annoying developments of the internet age is a weird resurgence of belief in a flat Earth. There has always been a tiny contingent of flat-Earth partisans, but thanks to modern technology, they can more easily find each other and produce slick videos that exploit the algorithms of YouTube and other social media platforms to reach gullible people who in a less connected age might never have encountered this community. They even get the occasional celebrity quasi-endorsement, as when NBA player Kyrie Irving publicly questioned whether the earth was round. (Irving later walked this back, and apologized to teachers whose job was made more difficult by his comments.)

The ancient measurement made by Eratosthenes, together with all details of the sun's daily and seasonal motion, shows how little question there really should be regarding the shape of the earth. This is not a subtle phenomenon that requires sophisticated equipment to tease out. The roundness of the earth is obvious, and proving it requires nothing more than measuring a couple of shadows.

In a similar vein, Newgrange's 5,000+ years of operation, combined with its massive scale, often fuels claims that it would've required more sophisticated science and technology than could be found in Ireland in 3000 BCE. This has taken many forms over the years (anti-Irish prejudice led many commentators in the 1700s and 1800s to attribute its design and construction to a dizzying array of other cultures, from the Romans and Danes to people from Egypt and even India), but the most recent form invokes "ancient aliens"—supposed extraterrestrial visitors providing advanced knowledge of astronomy to Neolithic humans for some inscrutable reason.

In fact, the technology needed to design the passage at Newgrange is shockingly basic; as Neil deGrasse Tyson memorably put it, all it takes is a stick in the mud.* Well, two sticks: one to mark an

---

* Tyson's essay, "Stick-in-the-Mud Astronomy," can be found online, and is well worth reading: https://www.haydenplanetarium.org/tyson/essays/2003-03-stick-in-the-mud-astronomy.php.

observation point, the second some distance away to sight on the rising sun. Tracking the position of sunrise through the course of a few years in this manner is enough to clearly define the line to the solstice sunrise, and thus the orientation of the passage. Then it's just a matter of coordinating work teams to move heavy rocks into position, which is tedious but not as technologically demanding as sometimes claimed.

Like its compeers, Newgrange stands as a testimony to the centrality of science to the human experience (certainly not as evidence of alien meddling in the human past). Its builders left no written records, so we know very little about them, but their monumental legacy unambiguously shows that they were doing science. They made careful and patient observations of the motion of the sun, constructed a model of the world that projected their observations into the future, and refined and shared that model over the course of decades, if not centuries. All of that scientific effort culminated in the magnificent solstice marker that still functions perfectly all these centuries later.

The nature of Newgrange is also evidence of the human preoccupation with time and timekeeping. It's no accident that all this scientific activity was directed toward the construction of a clock. In a Neolithic culture dependent on the seasonal cycles of plants and animals, the ability to mark and predict the turning of the year would be a matter of life and death. It's no surprise, then, that some of our most ancient evidence of science is dedicated to keeping time.

Chapter 2

THE SUN, THE MOON, AND THE STARS

Our first child was born in August 2008, and a few months before that my wife and I investigated the daycare options in the Schenectady area. One of the top recommendations, and our eventual choice, was the early childhood education program at the Schenectady Jewish Community Center. When we made our first visit, the director reassured us that it wasn't a problem that neither of us were Jewish—around half of the families with kids in the daycare aren't—but she also took pains to remind us that the program would be observing Jewish religious holidays as well as the regular collection of US civic holidays.*

We didn't entirely appreciate what that meant, having an admittedly hazy idea of a few major holidays (Rosh Hashanah, Yom Kippur, Hanukkah), until the fall when we discovered the Days of Awe cluster of high holy days. Between them, they accounted for seven days of daycare closures. What's more, as we learned in the following years, they don't occur at regularly predictable dates on the civil calendar; the first of this fall cluster of holidays, Rosh Hashanah, can fall anywhere between September 5 and October 6. We were caught off guard

---

* She also offered a very memorable description: "All our holidays are the same: somebody hated us, they tried to kill us, and it didn't work. Now it's a holiday."

those first couple of years, and had to scramble to find alternative childcare for those days.

That period of confusion was a vivid reminder that while it may *seem* universal, the modern civil calendar of (mostly) fixed-length months occurring in a rigid regular sequence is not, in fact, the only way to organize the passage of time. The Jewish holidays slide around relative to the US civil calendar because they're based on the Hebrew calendar, which has a different set of organizing principles reflecting different priorities.

We were lucky in that the Hebrew calendar does, at least, restrict the range of variation in a way that keeps all the daycare-closing holidays within the fall months. The other religiously based calendar in widespread use is the Islamic calendar, whose major holidays move through the entire civil calendar. They can even be celebrated twice in the same civil-calendar year. Again, it's a different sort of calendar, reflecting a different set of choices about what to prioritize.

All three of these calendars—the Hebrew calendar, the Islamic calendar, and the Gregorian civil calendar used in the United States—are shaped by thousands of years of theology and politics, but they're ultimately rooted in astronomy. The different systems reflect various ways of trying to accommodate the fundamentally incompatible cycles of the moving lights we see in the sky.

The seasonal motion of the sun across the horizon, and the corresponding variation in the length of a day, is the most consequential cycle visible in the apparent motion of objects in the sky. It's far from the only such cycle, though, because the sun is far from the only visible object in the sky.

The second most obvious celestial cycle is that of the moon, which also tracks across the sky in a variable but regular pattern, with the times of its rising and setting slowly shifting from day to day. In addition to the change in moonrise and moonset, the moon also appears to change shape. Beginning with the complete circle of the full moon, which rises in the east exactly at the moment the sun sets in the west, the moon rises later each night and also shrinks, to a half circle, then a waning crescent. At the time of the "new moon," moonrise and sunrise coincide, and the moon is not visible in the night sky. Then the moon returns as a thin crescent visible just after sunset, with its setting time moving later as the shape waxes to a fuller crescent, a half circle, and eventually to the complete circle of the next full moon, which sets exactly at the moment the sun rises.

The moon is the second-brightest object regularly appearing in the sky, and its cycle of phases is short enough to be readily apparent after only a few nights, making it an obvious choice for an astronomical timekeeper. The moon's motion across the sky is accompanied by an enormous number of visible stars, and these, too, present a pattern that changes over time. A person looking at the brightest visible stars will almost automatically perceive them as constellations, small groups forming particular patterns. Over the course of the year, the rising and setting times of any individual star or constellation will shift slightly earlier every day.

In the European tradition, this cycle of stars is usually described in terms of the apparent motion of the sun against the background of stars. This may seem a little counterintuitive, given that you can't see the stars when the sun is up, but as the year goes along, the particular group of stars visible on the horizon just before the sun rises or just after it sets changes, and these define the position of the sun. There are 12 or 13 different constellations in which the sun can be found over the course of the year, corresponding to the "zodiac signs" that form the basis of European astrology.

So, between the sun, the moon, and the stars, humans looking at the motion of objects in the sky have a wealth of cycles and patterns to track. Every human culture we know of has drawn on these cycles, in slightly different combinations, to make tools for tracking the passage of time. Maybe the most important fact about all these cycles, though, is that their lengths are different in such a way that none of them is an exact multiple of any of the others: a year is not an even number of solar days, nor is a lunar month, and a year is not an even number of lunar months. A 12-month calendar based only on the moon is almost 11 days shorter than a solar year, so a month that starts in the spring one year will begin in the winter within a decade.

The fundamental incompatibility of the duration of the lunar cycle and the length of a year means that a "perfect" calendar, in which the solstices and equinoxes always fall on the exact same dates, synchronized with the phases of the moon, is impossible to construct! That's never stopped people from trying, though, and looking for a scheme that meshes these cycles in a satisfying way has driven an enormous amount of innovation. In the next pages, we'll delve into the fascinating science behind the patterns we see in the motion of the sun and the moon against the background of the stars, and how these patterns change. We'll also trace some of the history of attempts to make the lunar and solar cycles fit together, and how these led to the different calendar systems we see in action today.

THE FAULT IS NOT IN OUR STARS

The cycle of the zodiac, like the seasonal motion of the rising and setting sun, has its origin in the orbital motion of the earth. As the earth goes around the sun, which stars we see just before the sun rises or just after it sets changes because our vantage point is changing. By the time the earth has rotated back around to bring a particular star directly overhead again, the planet has moved in its orbit by just enough to change the angle slightly. The resulting "sidereal day" (the time for a star to return to the same position in the sky) is about four minutes shorter than a solar day, and as a result the sun's position seems to shift relative to the distant stars. You see something of the same effect when looking out the window of a moving train: a distant feature of the landscape, such as a range of mountains on the horizon, will seem to remain more or less where it is, while closer objects like buildings pass by more rapidly. With a small amount of mental effort, you can choose to see this as if you and the mountains are standing still, while the buildings move past like the scrolling backdrop in a cheap cartoon.*

This apparent motion of the sun provides an alternative to tracking the position of sunrise or sunset to follow the progress of time through the course of a year. Stellar timekeeping has the great advantage of being portable: to accurately track the position of the rising sun on the horizon requires familiar landmarks observed from a consistent position, but the motion of the sun through the stars is the same for everyone on Earth. Tracking the rising and setting positions of the sun is well suited to monumental architecture, but knowing which stars are visible just after sunset in a given season is much more useful to wandering groups, or civilizations that spread out over wider ranges.

---

* The positions of the stars are not, in fact, perfectly fixed, but change very slightly as the earth orbits the sun. Sufficiently careful measurements of the apparent position of a relatively nearby star will show it tracing out a tiny circle, but the distance to even the nearest stars is so enormous that this "stellar parallax" is extraordinarily difficult to detect, and it was not successfully measured until the early 1800s, after centuries of effort.

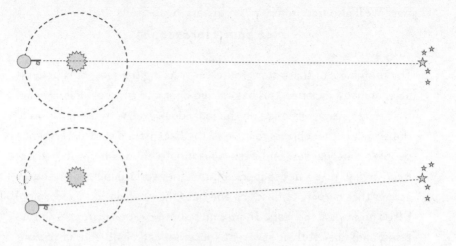

The difference between solar and sidereal days: as the earth moves in its orbit, the
time needed to rotate to bring a distant star back to directly overhead (a sidereal
day) is less than the time needed to rotate so the sun is directly overhead again.

For this reason, basically every ancient civilization has its own body of
knowledge about the stars, usually encoded in myths and legends connecting
particular stars to particular seasons. Many of these stories are about Sirius, the
brightest star in the night sky.

Several thousand years ago in northern Africa, Sirius would disappear
behind the sun in May, then return to the night sky in mid-July, rising above
the eastern horizon just before dawn. This "heliacal rising" occurs near the start
of the period when the Nile River floods the northern regions of its course, a
crucial event for agriculture in Egypt.* Accordingly, the star became associated
with the Egyptian goddess of the flood, Sopdet, and later with fertility more
generally.

Across the Mediterranean in Europe (and in most of North America), the
period following the reappearance of Sirius corresponds to the height of summer, so its heliacal rising became associated with heat and drought, a not particularly positive omen. The Greeks and Romans placed it in the constellation

---

* The exact date of heliacal rising changes with latitude, and slowly changes over time due to
a gradual shift in the axis of the earth's rotation and the orientation of its orbit. The modern
dates of Sirius's heliacal rising are several days later relative to the seasons than they were in
ancient times.

Not Your Horoscope

The usefulness of these star groups for tracking the seasons is undeniable. In many societies, this has carried over into a more dubious belief that they have significance on the individual level, with the position of the sun and other objects relative to the background stars determining people's personalities, and even appropriate actions on a daily basis. Even today, many newspapers around the world publish daily horoscopes that purport to describe the best things to do (or avoid doing) based on reading the stars. There's no scientific rationale for these systems, and most of their apparent successes can be attributed to confirmation bias. In his 1980 book *Flim-Flam!*, magician-turned-activist James Randi recounts an episode from his youth in which he wrote an astrology column under the name "Zo-Ran." Rather than casting any actual horoscopes, though, he copied daily readings from other astrology columns, assigning them to signs and dates at random. To his dismay, readers of the paper found Zo-Ran's supposed predictions "right smack on," and eagerly awaited his next round. The reason is simple psychology: people remembered the predictions that "came true" and forgot or rationalized away those that didn't.

Canis Major, the dog,* which is the root of the idiom "dog days of summer," describing the unpleasantly hot stretch that often occurs in late July and early August. Meanwhile, for the Maori in New Zealand and various Polynesian civilizations in the southern hemisphere, where the seasons are reversed, Sirius was associated with winter.

In addition to the significance attached to the appearance and disappearance of individual bright stars, most cultures have attached significance to broader *groups* of stars, using them as touchstones to divide the year into seasons depending on which groups of stars are closest to the rising or setting sun. Starting with the Babylonians, most civilizations in the western part of Eurasia

---

* The classical poet Homer refers to Sirius as "Orion's Dog" in one passage of the *Iliad*.

have defined 12 or 13 constellations from the stars in the band around the sun's apparent path through the sky; each of these corresponds to roughly 30 degrees of the full 360-degree orbit.* The names and properties associated with these constellations tend to reflect the seasons in which their heliacal rising and setting occur, with folklore associating appropriate agrarian activities with each.

THERE'S NO DARK SIDE OF THE MOON, REALLY; MATTER OF FACT, IT'S ALL DARK

The nearly universal choice to divide the year into 12 parts may seem a little unusual, but it's probably no accident that most cultures attuned to the stars as a way to track the sun's motion chose that number. One-twelfth of a "tropical year"—defined as the time from one March equinox to the next—is a bit more than 30 days, while the cycle of phases of the moon, from the first visible crescent of one cycle to the next, takes 29.53 days on average, a close match. This is a very convenient span of time, with the moon's shape changing visibly over just a couple of days, so nearly every culture has used the moon to track time.

The changing shape is a result of the way both the earth and the moon are illuminated by light from the sun. Each body has half of its surface lit at all times, but as the moon orbits the earth, the angle from which we see it changes, changing the apparent shape. When all three objects are in line with the moon between the earth and sun, the moon does not appear in the night sky at all, and only the unlit side would be visible from Earth if it weren't hidden by the glare of the sun.† As the moon moves around the earth, we begin to see the lit side, first as a thin crescent visible just after sunset, then as a half circle, when the line between the earth and the moon is at right angles to the line between the earth and the sun.

---

* The traditional Chinese scheme subdivided 12 named months into 24 solar terms, each covering about 15 degrees of the sun's apparent motion.
† During this time, the far side of the moon, which never faces the earth, is fully lit, which is why it's a mistake to refer to "The Dark Side of the Moon" unless you're talking about Pink Floyd.

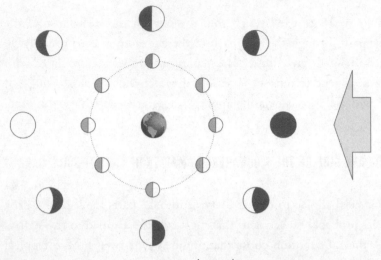

Lunar phases.

The full moon occurs when all three objects are in line again, but with the earth between the sun and the moon.* As the moon's orbit continues, the amount of the lit portion visible from Earth shrinks again, to a half circle, then a thin crescent visible just before sunrise, then the moon is hidden in the sun's glare again and the cycle repeats.

As the moon goes around the earth, the angle from which those of us on Earth see it not only changes what fraction of its lit surface we can view, but also its apparent position against the background stars. The moon's path through the stars is nearly the same as that of the sun, just faster: the moon moves through the constellations of the zodiac once a month, while the sun requires an entire year.† In fact, the moon's orbital motion is rapid enough that careful observations can show it changing positions through the course of a single night. The

---

* The third straight-line arrangement, with the sun between the earth and the moon, never happens, which is a good thing because that would have apocalyptic consequences. When the conditions are just right, the two straight-line configurations that do happen lead to eclipses of the sun or moon—which are spectacular, but not devastating. We'll talk more about the orbit of the moon in Chapter 9.

† As with the motion of the sun, there are numerous ways this movement of the moon through constellations has been used for divination. Even in 2021, you can find gardening websites offering earnest advice as to the best dates for planting particular kinds of vegetables and flowers based on the zodiacal location of the moon. It should, but sadly doesn't, go without saying that there is no plausible scientific basis for this.

moon appears to move against the stars by about one moon diameter (half a degree) per hour, which means that between rising and setting, the full moon will move by about one-fifth the width of a zodiac sign on the sky. This is critically important to the use of the moon for navigation (as we will discuss in Chapter 7).

Incommensurate Cycles

It's not difficult to imagine a simpler world in which the cycle of the moon's phases and the cycle of the seasons mesh perfectly. If the moon orbited the earth a little more slowly, taking exactly 30 days, and the earth orbited the sun a little more quickly, taking only 360 days, the combination of the sunrise position and the phase of the moon would unambiguously identify a single day within the year, and celestial timekeeping would be a breeze. There would be exactly 12 lunar cycles per solar year, and nobody would ever need to buy a new calendar, or mentally recite that "Thirty days hath September . . ." rhyme to remember the length of the current month.

This, though, is not the world in which we live. A tropical year is 365.24217 days. The average length of a "synodic month," defined as the time from one new moon to the next, is 29.530588 days.* These are not simple multiples of one another—there are a bit more than 12.37 synodic months in a tropical year, meaning that from year to year the phases of the moon are not at all synchronized with the seasons of the year.

The incommensurability of these cycles means that, when designing a calendar, something's got to give: at least one of these cycles will not fit perfectly. Given the importance of the sun's motion for driving the seasons, it might seem logical, particularly for an agrarian society, to simply disregard the moon. The sun's motion along the horizon is very slow, though, with day-to-day changes that are barely perceptible, while the moon's phases are much more in keeping with a typical human attention span. The temptation to design some scheme for tracking time by the moon is all but irresistible.

---

* The large number of decimal places in these, which amounts to defining a year with millisecond precision, is a testament to the precision of timekeeping developed through centuries of observation, a point we'll come back to many times in the coming pages.

The most straightforward way to track time using the moon is to simply ignore everything else. This is best illustrated by the Islamic calendar, which contains 12 strictly lunar months of 29 or 30 days each (the start of a new month is officially announced on the sighting of the crescent moon just after sunset). This leads to a 12-month year of around 355 days, which is shorter than the tropical year by 10 or 11 days.

As a result, the holy days of Islam drift relative to the seasons: for example, the month of Ramadan, during which devout Muslims fast during daylight hours, started around April 24 in 2020, about a month after the spring equinox. In 2010, Ramadan began on August 11, about a month before the fall equinox, and in 2030 it will begin around January 6, a few weeks after the winter solstice.*

A former student who was a devout Muslim described his experience of this drift in a way that made it very concrete. He noted that at the time he became old enough to be expected to observe the sunrise-to-sunset fast, Ramadan was in the winter, when the days are short, so the fast was relatively easy. As he got older, the start date slipped backward relative to the seasons, and by the time he was in college, Ramadan was starting in late summer. "Every year, it gets a little harder," he said.

Needless to say, the Islamic calendar is difficult to use for agricultural purposes, so the countries that follow it must maintain some other system for determining planting and harvest dates. Many societies employ a trick known as "intercalation" to keep their mostly lunar calendar in synch with the seasons: a typical year is 12 lunar months, but every now and then an extra month is added.

The most elaborate example of an intercalary scheme is the Hebrew calendar used to determine Jewish holidays and complicate our childcare arrangements (and as a civil calendar in Israel). The base of the calendar is a set of 12 lunar months of either 29 or 30 days (which on average provide a reasonable approximation of the 29.53-day lunar month), with occasional years lengthened or

---

* The year 2030 is an interesting example of a Gregorian year in which Ramadan will start twice, first in January and again in late December.

shortened by one day to ensure that Rosh Hashanah (the start of the next year) falls on a Monday, Tuesday, Thursday, or Saturday.*

The resulting calendar has a year of either 353, 354, or 355 days, which, like the Islamic calendar, falls 10 to 12 days short of the tropical year, and would lead to the major Jewish holidays slipping relative to the seasons. As the Torah associates particular holidays with particular seasons, though, this drift would pose a theological problem; thus, the calendar allows for the insertion of an extra 30-day month every two or three years,[†] making up for the missing days and keeping the high holidays at the appropriate points in the cycle of seasons. The longer years are subject to the same rules regarding the day of Rosh Hashanah as ordinary years, so the Hebrew calendar ultimately has *six* possible year lengths: 353, 354, 355, and 383, 384, or 385 days.

Thousands of years ago, the start date for each month was determined by the sighting of the new moon, and the intercalations were determined by rabbis in Israel. As the Jewish population spread over distances too great to easily communicate these decisions, particularly after the destruction of Jerusalem by the Roman Empire in 70 CE, a gradual shift was made to a calculated calendar; the formal rules in use today were established by 1178 CE (in the Gregorian calendar). The rules prescribe seven leap years out of every 19,[‡] in a fixed cycle: years 3, 6, 8, 11, 14, 17, and 19. This 19-year cycle is known as the "Metonic cycle" after the Greek astronomer Meton of Athens, who introduced it to the Greek calendar in 432 BCE, though like almost every other accomplishment we attribute to Greek mathematicians and astronomers, it was used by the Babylonians before that. The pattern of intercalations is further modified by the Rosh

---

* This is related to the Shabbat rules prohibiting work, which also apply to Yom Kippur. If Yom Kippur were to fall on the day after Shabbat, the preparations necessary for its observations would be forbidden, and likewise if it were to precede Shabbat by one day, the Shabbat preparations would be forbidden. To avoid these complications, the length of a year is adjusted as needed to ensure that Rosh Hashanah falls on one of the four permissible days.

† The extra month gets the same name as the final month of the regular calendar, so a leap year ends with "Adar I" and "Adar II." The inserted month is taken to be the first of these, so the holiday of Purim is celebrated in Adar II during a "Shana Me'uberet" or "pregnant year."

‡ Rounded to the nearest day, 19 tropical years and 235 synodic months are both 6,940 days, and 235 months divides into 19 years of 12 months with 7 months left over.

Hashanah postponement rules, meaning that a given sequence of years in the Hebrew calendar won't repeat exactly for 689,742 years.*

The resulting calendar is extremely complex—every time I've taught my college course on the history of timekeeping, I've had at least one student attempt to write a computer program to convert future dates on the Gregorian calendar to their Hebrew calendar equivalents, and none have ever gotten it completely correct. The end result is very effective, though, in terms of meshing the lunar and solar cycles: on average, the Hebrew calendar loses one day relative to the seasons every 217 years. Presumably this could be fixed by an additional intercalation at some point in the distant future, but we're a long way from needing to make that change.

Forget the Moon

Yet a third approach to dealing with the incompatible cycles of the sun and moon is to simply forget about trying to synchronize the calendar with the phases of the moon. This tactic has a very long history and ultimately forms the basis of the modern civil calendar.

The earliest such calendar is the civil calendar of ancient Egypt. It's difficult to pin down exactly when the Egyptian civil calendar was created; some sources claim it stretches back to 4242 BCE, but that predates archaeological evidence of civilization in the area. A more plausible start date is closer to 2782 BCE, established by working backward from the structure of the calendar and its tie to the rising of the star Sirius.

As mentioned above, the most important event in the agricultural year in ancient Egypt was the annual flooding of the Nile, which typically begins right around the time of the heliacal rising of Sirius. The civil calendar established in ancient Egypt placed the start of a new year at this rising, beginning a cycle of 12 months of 30 days each, with the months grouped into three seasons (the seasonal names translate approximately to "Flood," "Emergence," and "Harvest"). This accounts for 360 days, a bit more than five days short of the tropical year; this difference was accounted for by the addition of five extra holidays, not part of any month, celebrated as the "birthdays" of five major gods.

---

* We should be so lucky . . .

The resulting 365-day calendar is impressively accurate for 2800 BCE, but still too short by a bit less than a quarter of a day. As a result, the heliacal rising of Sirius would slip relative to the civil calendar by roughly one day every four years. So, for example, if Sirius rose just before sunrise on the first day of the first flood month in 2782 BCE, that rising would fall on the *second* day of the first flood month in 2778 BCE, and the *third* day of the first flood month of 2774 BCE, and so on.

In an age before telescopes, the precise date of the rising of a particular star is sufficiently subjective that it would take a good number of years before the slippage became undeniable. This could be fixed by the addition of an intercalary day from time to time, but the Egyptians opted to simply document the civil date of the heliacal rising of Sirius as it moved through the calendar.

It takes more than 1,400 years for the rising of Sirius to come all the way back around to the first day of the first month of the civil calendar, but the remarkable longevity and stability of Egyptian civilization allowed for the identification of this "Sothic cycle." In fact, this is what allows us to estimate the start date of the Egyptian civil calendar: the Roman writer Censorinus noted that the start of the calendar for a particular Egyptian year coincided with the rising of Sirius in what we would call 139 BCE. Tracing the Sothic cycle backward from that fixed point is one of the tools that allows archaeologists to make reasonably good identification of the modern-calendar equivalent of many key dates in Egyptian history. Assuming that the calendar was formally implemented in a year when Sirius rose on the "correct" date gives a probable start of 2782 BCE.*

The 30-day months of the Egyptian civil calendar probably reflect the influence of an earlier lunar system but represent a notable innovation in the history of calendar development in that they are no longer strictly tied to observations of the moon. The 365-day cycle of strictly defined months makes for a simpler calendar, in which dates can be determined by merely counting days, which has some significant advantages for an empire coordinating activities over a large geographic area. This probably accounts for the adoption of fixed-term months and readily calculable dates in most later calendars.

---

* There is, of course, a little slipperiness to this date, depending on the exact length used for the Sothic cycle and the reliability of Censorinus's information.

On the other hand, the slow drift of the Sothic cycle decouples the civil calendar from the agricultural year. In the case of Egypt, the heliacal rising of Sirius and the annual flooding of the Nile provide an easy check on this, which may explain why even though the Egyptians were aware that the calendrical drift could be corrected by intercalation of days, they never implemented this correction. The 365-day Egyptian civil calendar remained in use with minimal modification for almost 2,800 years, until change was forced on them from the outside, by the next generation of calendrical innovators.

Across the Mediterranean, 2,000 Years Later

The traditional Roman system dated years "*ab urbe condita*," based on the number of years elapsed since the historical founding of the city of Rome in what we now call 753 BCE. Like most ancient societies, the Roman calendar had its origin in a system of lunar months that would come up well short of a tropical year. The Roman religion, with its diverse array of gods, also required numerous rituals to coincide with particular seasons of the year, so as with the Hebrew calendar, occasional intercalary months were used to keep the official calendar in synch with the solar cycle.

In addition to being very religious, the Romans were also a highly political people. The decision of when to add an intercalary month was made by the *pontifex maximus*, which was a political office as well as a religious one. With depressing inevitability, this power was abused: extra months were added in years when doing so would extend the term in office of politicians aligned with the pontifex maximus, or left out to shorten the terms of their opponents. By the middle of the first century BCE, when Julius Caesar completed his rise to power, the Roman calendar was badly in need of reform. One of the final steps in Caesar's ascent was the conquest of Egypt, where he was famously captivated by Cleopatra.* He ended up spending several years in Egypt before returning to Rome, so it's probably no accident that when he finally imposed a new calendrical system, it took much of its structure from the Egyptian civil calendar.

Caesar's calendar began with 12 months having fixed lengths, alternating between 30 and 31 days. This gives a standard year 366 days, a bit too long, so

---

* Formally Cleopatra VII Philopator, the last queen of the independent Ptolemaic kingdom of Egypt before it was absorbed by Rome.

the month of February was shortened by one day (from 30 to 29). From the experience of the Egyptians, though, this was known to be too short by about a quarter of a day, which led to Caesar's bold innovation, the progenitor of the modern system of "leap years." Every fourth year, an extra day would be added to February, bringing the average year length to 365.25 days and correcting for the drift of the Sothic cycle.

The initial introduction of the Julian system required one "*ultimus annus confusionis*" to bring the calendar back into synch with the seasons: this "final year of confusion" (46 BCE in the modern system) ran to an amazing 445 days. The Julian calendar was officially introduced in 45 BCE, and after a small correction to the leap year implementation under the emperor Augustus in 8 BCE,* it was the official calendar for the remainder of the Roman empire and into the post-Roman era of Europe. This brought a much-needed sense of stability to the seasons: important astronomical and agricultural events were synchronized with the civil months, and religious rituals occurred in the proper seasons.

NOT A FLAT CIRCLE, BUT A SOCIAL CONSTRUCT

The three major calendars described here present three different choices about how to reconcile the two most significant cycles we see in the sky, namely the motion of the sun and the phases of the moon. These reflect different priorities regarding which of the two is more important: the Islamic calendar gives highest priority to the phases of the moon, allowing the months of the year to drift through the seasons. The Julian calendar (immediate ancestor of the modern Gregorian civil calendar) gives the highest priority to the seasons of the year, ensuring that spring and summer always start in the same months, but losing any direct connection to the phases of the moon. The Hebrew calendar represents an attempt to strike a balance between the two, ensuring that the holy days always fall at the appropriate point in the lunar cycle, but inserting the occasional month to keep them in the right general season as well.

---

* At which time the months of Quintilius and Sextilius were renamed in honor of the Caesars, the origin of the modern names "July" and "August"; February was also shortened by another day to ensure that Augustus's month was the same length as Julius's, leading to the confusing modern mishmash of month lengths.

The (uneasy) coexistence of these three calendars illustrates the important point that time is a social construct. The orbits of the earth and the moon are empirical facts of astronomy, but calendars are made by humans and reflect the particular interests and priorities in the societies that create them. While rooted in astronomy, the calendar systems were developed through a process of compromises between theological, agricultural, and political interests. The full range of these interests can be seen in action in the relatively well-documented transition from the Julian calendar introduced by the caesars to the Gregorian calendar that is now the international standard. This took place in Europe in the late 1500s, against the backdrop of the Protestant Reformation, and as we'll see in the next chapter, that break from long-established papal authority made what is in fact a rather small tweak to the Julian calendar rules into a contentious matter of theology and international politics. The change also led to a surprising gap in the calendar a century and a half later.

Chapter 3

"GIVE US OUR ELEVEN DAYS!"

I n 1755, the English painter William Hogarth produced a series of satirical oil paintings collectively titled "Humours of an Election," depicting the events around a parliamentary campaign.* The first of these, "An Election Entertainment," shows an assortment of unappealing Whigs gathered in a tavern while a riot rages outside. One of the many small details crowding the frame is a trampled placard, presumably claimed as spoils by one of the bludgeon men battling in the riot. The slogan on the placard is "Give Us Our Eleven Days!"

Hogarth's paintings were modeled in part on the elections held in Oxfordshire in 1754, which were particularly contentious, and did, in fact, involve some wrangling over a "lost" period of 11 days. In 1752 in the United Kingdom, September 2 was by decree of Parliament followed by September 14. The dropping of those 11 days led to some predictable confusion about the scheduling of events and some disputes over sharp dealing—landlords asking a full month's rent for a period of just 19 days, or employers withholding wages for the missing days. These disruptions were still fresh in the minds of Tory voters who saw the change as a Whig imposition.

---

* These are displayed in Sir John Soane's Museum in London, along with many other Hogarth works. The museum was originally Soane's house and is well worth checking out for the many unique architectural features he designed to display his art collection.

The dropping of 11 days was required to bring Britain (and its colonies) into synchrony with the Gregorian calendar that had already been in use in continental Europe for a century and a half. Happily, Hogarth's satirical painting aside, there seem not to have been actual riots over the end of "Old Style" dating; contemporary newspapers and diaries record that the calendar shift went about as smoothly as could be hoped for such a far-reaching change. Within a few years, the dislocations caused by the date change were essentially forgotten, save when painters and Whig historians wanted a colorful anecdote about the ignorance of the common people. "Give Us [Back] Our Eleven Days" had a longer life as a slur against rural rubes than an actual political slogan.

However limited its political effect, the calendar reform in Britain is a fascinating event in its own right. The act of Parliament that shortened 1752 by 11 days was the last small piece of a complex story that began centuries earlier, with a theological problem regarding the date of Easter.

AN ALL-TOO-MOVABLE FEAST

As noted, the Julian calendar was the official calendar for the entire imperial period of Rome and well into the post-Roman era of Europe. It struck a nice balance between ritual and bureaucratic interests, with fixed-length months and occasional leap years serving to keep the civil calendar in synch with the seasons of the year, and allowing the important seasonal festivals to fall on readily predictable dates.

For the most part, anyway. As was the case with the Egyptian civil calendar, the enormously large span of time over which the Julian calendar was used allowed tiny errors to accumulate to the point where they became significant. The 365.25-day Julian year (with one day added every four years) is longer than the 365.2422 days of the tropical year by a bit more than 11 minutes. That difference isn't noticeable on the scale of a year or two, but over time, those minutes add up, causing the calendar to slip relative to the seasons by about one day every 128 years. By 725 BCE, when the Venerable Bede was writing his treatise on the calendar in a monastery in Northumbria, the difference between the actual equinox and the date when it was supposed to fall according to the Julian calendar was nearly a week. That difference is readily measurable with a sundial, even in the cloudy climate of Bede's British Isles, at least if you're a monk with limited entertainment options.

A difference of six days between the predicted and actual date of the spring equinox is measurable, with some care, but might not seem terribly significant. Even at the time, this probably wasn't exceedingly important to the vast majority of people—practically speaking, the vagaries of weather shift the date of first planting by more than that from year to year, so most farmers would hardly notice. For an astronomically inclined eighth-century Christian monk, though, this was enormously significant.

Christianity began as an offshoot of Judaism, and then grew to become the state religion of the late Roman Empire. Given these origins, it's no surprise that medieval Christianity inherited a strong preoccupation with making sure that major rituals take place in the proper seasons. Easter, the most important feast in the Christian calendar, celebrates the events surrounding the death and resurrection of Jesus. The Gospels clearly describe these as taking place around the Jewish holiday of Passover, commemorating the escape from Egypt, which is celebrated on the fifteenth day of the month Nisan in the Hebrew calendar. The luni-solar nature of the Hebrew calendar guarantees 15 Nisan will take place on a full moon in the spring, but the exact date is not strictly tied to the motion of the sun. Depending on the cycle of intercalations, it may be either the first or second full moon following the March equinox. But since the Julian calendar has no direct tie to the phases of the moon, calculating the date of Easter posed a serious challenge.

A seemingly obvious solution to this problem would be to simply celebrate Easter on 15 Nisan or the Sunday closest to 15 Nisan, and indeed, this was a common practice in the early Christian church. There are some theological problems, though, with having the date of your most important religious festival tied to the calendar of a rival sect. Worse yet, the lunar nature of the Hebrew calendar and the lack of a central authority in the early centuries of the Jewish diaspora meant that local Jewish authorities in different parts of the empire didn't necessarily agree what month it was. As a result, early Christian theologians expended a great deal of effort devising a method for calculating the date of Easter that would bring order to this calendrical chaos.

The proper procedure for celebrating Easter was one of the many topics debated by the first Council of Nicaea, convened by the Emperor Constantine in 325 CE. While the council stopped short of formally endorsing any particular method for determining the date of Easter,* they declared that Easter should be celebrated on a single day by all Christians worldwide. Given fourth-century logistics, this is a strong argument for a computation-based method of determining the date of Easter, and the next several centuries saw furious activity in this area.

The essential concept of the "*computus,*" as it evolved over several hundred years, is the establishment of a Christian version of Nisan 15, using a simulated lunar calendar. This involved setting up a "shadow" ecclesiastical calendar of months that alternate between 29 and 30 days and inserting occasional extra months in keeping with the 19-year Metonic cycle. Since the Christian authorities putting this together did not feel obliged to respect the Sabbath rules, the shadow calendar is simpler than the Hebrew calendar, with the start of each ecclesiastical month falling on a calculable date in the Julian civil calendar.

The sole purpose of this shadow calendar is to predict a date for each full moon, which is held to fall on the fifteenth of the relevant ecclesiastical month. Since the Julian calendar is tied to the tropical year, the spring equinox falls predictably on March 21, so the date of Easter is the first Sunday after the first ecclesiastical full moon after March 21. Simple, right?

In principle, the calculation of the date of Easter can be done by anyone with some basic mathematical ability and a list of the rules governing the ecclesiastical calendar; the Venerable Bede spent a considerable amount of time in one of his books explaining how it was done. Humans being kind of lazy, though, most Christians relied on computus tables worked out by individual clerics and disseminated widely. Two of the most famous sets of Easter tables were the Dionysian Tables calculated by Dionysius Exiguus in 525 CE and the Victorian Easter tables prepared by Victorius of Aquitaine in 457 CE. While this method is appealingly simple, the calculations were somewhat prone to error, so the precise date of Easter continued to be disputed well into the eighth century. In Bede's home kingdom of Northumbria, this famously led to absurdities like the "two Easter" year of 665 CE in which Queen Eanflæd, raised according to one tradition, was fasting as required on Palm Sunday on the same day her

---

* Or even declaring that it should always be celebrated on a Sunday, rather than the date corresponding to 15 Nisan.

husband, King Oswy, was celebrating the feast of Easter as calculated based on the other set of tables.

This was the political and theological context in which Bede made his observations of the solstices and equinoxes, and he found a clear difference between the nominal equinox date of March 21 in the Julian calendar and the actual astronomical equinox a few days later. The use of a fixed civil-calendar date for the equinox in the Easter computus made this shift a matter of some religious importance—even once they managed to settle on a single date for Easter, they might've been celebrating on the first Sunday after the *wrong* full moon! Given Bede's somewhat remote location and the general political instability of the time, though, several more centuries were to pass before any action was taken to fix the problem.

THE GREGORIAN REFORM

While determining the correct date for Easter was accepted as an important theological matter, the great political powers of Europe had their hands full with more pressing crises. The medieval period saw the rise of Islam in the Middle East and the ensuing Muslim conquest of North Africa and Spain; the launch of the Crusades to recapture the Holy Land; the Reconquista in Spain; innumerable religious schisms, including an extended period with multiple rival popes; and of course the Black Death wiping out more than a third of the population of Europe in the mid-1300s. In the face of all that, calendrical reform just couldn't get any traction.

The combination of slow but steady improvements in astronomy and mathematics with the slowly accumulating drift of the Julian calendar, though, kept the subject on the minds of the mathematically astute. Numerous writers noted the growing discrepancy between the Julian calendar and the astronomical equinox, including a pair of colorfully named Swiss monks, "Notker the Stammerer" around 900 CE and "Notker the Thick-Lipped" around 1000 as well as the polymath German monk known as Hermann the Lame around 1040. The English monk and protoscientist Roger Bacon noted in 1267 that the astronomical equinox was by this time taking place some nine days before the ecclesiastical one, and in his characteristically polemical style called the

Julian calendar "*intolerabilis, horribilis, et derisibilis*" ("intolerable, horrible, and laughable" in the scholarly Latin of the day).

This growing discrepancy eventually began to draw some official attention. Bacon suggested a reform of the leap-year system that would've dropped one day from the calendar every 125 years, but Pope Clement IV died before any action could be taken, and his successor was not interested in Bacon's ideas. Several attempts to find a solution were launched in the 1400s, the most serious in 1475, when Pope Sixtus IV summoned the famed astronomer and mathematician Regiomontanus (Johannes Müller von Königsberg) from Germany to Rome. Regiomontanus would have been an ideal person to develop an improved calendar, but he died soon after reaching Rome, and the work fell apart without him.

International and internal church politics thus delayed reform of the calendar for several centuries. There's some irony, then, in the fact that the eventual resolution of the issue came about as an offshoot of yet another political and religious schism.

In 1517, a German monk named Martin Luther published his 95 Theses, a list of grievances about corrupt practices within the Catholic Church (the oft-told tale of Luther nailing them to the church door is a later embellishment). Luther's writings kicked off the Protestant Reformation that splintered the Catholic Church into the wide range of Christian denominations we have today, and it was the trigger for more than a century of sectarian warfare. In response to Luther and other reformers, the Catholic Church launched a "Counter-Reformation" to clarify and formalize Catholic doctrine. The centerpiece of this effort was the Council of Trent, formally convened by Pope Paul III in late 1545. Thirty-seven years and six popes later, this led to the adoption of the modern Gregorian calendar.

The Council of Trent was active from 1545 to 1563, with three main sessions separated by years of inactivity due to wars and plagues and political squabbles within the church. In modern terms, the general tone might be characterized as a call for a return to "traditional values," reaffirming nearly all of the doctrines questioned by Luther and others. In the manner of committees throughout history, they also delegated tasks to various other bodies: one of the final decrees issued on the dissolution of the council by Pope Pius IV in 1563 concerned the work of such a subcommittee, which had been tasked to draw up a list of heretical books. The decree charged them to complete their work and present it to the pope—the infamous *Index Librorum Prohibitorum* of books forbidden to Catholics—and almost in passing directed the same group to comprehensively revise the missal and breviary, which spell out the Bible readings to be used in Masses through the course of the year.

Being essentially a schedule of seasonally appropriate services, revision of the breviary and missal was obviously closely connected to reform of the calendar. The revised breviary issued in 1568, shortly after the Council of Trent concluded its work, thus included a minor reform of the ecclesiastical calendar. This reform, however, which would have shifted the ecclesiastical date by four days and dropped a day from the calendar every 300 years thereafter, did nothing to address the mismatch between the civil calendar and the tropical year, and it was widely ignored.

This did not sit well with church leader Ugo Boncompagni, who had served as one of the many ecclesiastical lawyers working with the Council of Trent. After he took office as Pope Gregory XIII in 1572, he decided to settle the calendar issue once and for all. Gregory formed a commission to study the question and come up with a new calendar, but no record survives of the exact date of the formation of this commission, nor of its official members. The only official document we have concerning the commission is its final report, presented to the pope in 1581, introducing what is now known as the Gregorian calendar. The report is signed by nine people: a cardinal and a bishop from the Roman church, a patriarch of the Eastern church, a lawyer, a translator, a historian, and three scholars who dealt with the scientific aspects of the reform (the Dominican astronomer and mathematician Ignazio Danti, the Jesuit astronomer Christopher Clavius, and a doctor from Calabria in southern Italy named Antonio Lilius).

Antonio Lilius was in some sense both the least and the most important of the signatories to the report. He was there less for his own merits than as a stand-in for his brother, who had died in 1576. Aloisius Lilius,* also a physician in Calabria, had written a treatise on the calendar that, in the end, was adopted nearly wholesale by the commission and presented to the world. The

---

* His given name was Luigi Giglio (sometimes "Lilio," because spelling was more erratic in those days), but in formal contexts this was Latinized into "Aloisius Lilius."

astronomer Clavius was the most notable scientist of the signatories, followed by Danti, but their primary role was to confirm the validity of the reforms suggested by Lilius, whom Clavius praised as the *"primus auctor,"* the primary author of the reform.

One of the bigger obstacles to calendrical reform through the centuries had been the question of how to approximate the length of the tropical year in some manner that was simple and practical, along the lines of the Julian calendar's leap-year rule. Many of the proposals floated over the years were impractically complex, needing days added to or dropped from the calendar at inconvenient intervals. Lilius's scheme was a straightforward modification of the leap-year rule: dropping the extra day from three of every four century years. For example, a year whose number in the Common Era dating scheme ends in "00" is *not* a leap year, even though it ordinarily would be, unless that year is also divisible by 400: thus, 1900 was not a leap year, but 2000 was. This shortens the average calendrical year by 3/400ths of a day from its Julian value of 365.25 days to 365.2425 days. This still isn't a perfect match to the tropical year, being too long by about 3/10,000ths of a day (just under 26 seconds), but it's a much better approximation, and more importantly an easy-to-remember rule.

Of course, the most pressing question facing the commission from the standpoint of the church was the theological one: how to fix the date of Easter in a way that kept it in the proper position relative to both the seasons of the year and the phase of the moon. While some astronomers—including Clavius, at least initially—argued for basing this on direct observations of the moon, the church preferred a solution that did not require astronomical expertise on the part of the clergy. Lilius found an elegant solution to this problem as well, based on reforming the ecclesiastical calendar in a manner similar to the change to the leap-year rule. Under Lilius's scheme, every 300 years the church shortens the length of its shadow lunar calendar year by a day, and after seven such cycles, they wait 400 years before dropping the next day. Then the 2,500-year cycle starts again.*

Being the work of a committee, of course, the Gregorian reform also included some inelegant aspects resulting from political compromises, mostly

---

* In theory, at least—we're only 400 years into the Gregorian calendar era, so not all of the rules have yet come into play.

concerning how to address the accumulated errors of the Julian calendar. While the "traditional values" ethos of the Council of Trent might've argued for shifting the date of the equinox back to March 14 (its position in the era of the caesars), the commission instead opted for shifting it to March 21. This matched the date used by the Council of Nicaea, but more importantly avoided the need to expensively reprint thousands of breviaries that had already been distributed. This shift was accomplished by the same technique later used in Britain: dropping 10 days from the calendar for one year.* Thursday, October 4, 1582, was followed by Friday, October 15, with the specific dates to be skipped chosen because no significant feasts or saints' days fell in that span.

Implementation of the Gregorian Calendar

The calendar reform was announced in the papal bull "*Inter gravissimus*,"† dated February 24, 1582, ordering its implementation that fall. As a reward for his late brother's work on the calendar, Antonio Lilius was granted the exclusive right to publish calendars using the new system for a period of 10 years. By that September, though, with the critical day fast approaching, Lilius was unable to meet the demand, and the grant was rescinded to allow for a dramatic increase in the production capacity.

The Reformation that indirectly triggered the development of the Gregorian calendar also complicated its adoption. Only the staunchly Catholic monarchies in Spain and France and the Polish-Lithuanian Commonwealth adopted the new system more or less on schedule. Many Protestant and Eastern Orthodox nations were skeptical of anything emanating from Rome and, at least initially, refused to have anything to do with it. Over the next century, the holdouts slowly gave in, as the demands of international trade brought most of continental Europe in line. Some Protestant denominations maintained independent calculations of the Easter date, but this was mostly a nominal change,

---

* Lilius's original proposal called for shortening 10 years by one day each, but Clavius argued that it would be better done in a single step.

† Like all papal pronouncements, this takes its name from the first two (Latin) words of the document, which starts out: "Among our serious pastoral duties, not the last is that we care to complete those sacred rites reserved by the Council of Trent, with the guiding assistance of God."

as they tended to adopt formal rules that produced the same result as the Gregorian method, just without reference to the pope.

Unusually for a Protestant nation, Britain under Queen Elizabeth I came tantalizingly close to reforming their calendar in the 1580s. Elizabeth's court astrologer John Dee was charged to evaluate the new system. He produced a lengthy report analyzing the scientific reasons behind the change, as well as recommending that the nation should adopt a new calendar more or less equivalent to the Gregorian system,* complete with rationales rooted in distinctly British interests. The queen and her privy council were persuaded by Dee's arguments and sought to move ahead, but the archbishop of Canterbury balked, and the matter was dropped. Two more attempts to reform the calendar, one around the time of Oliver Cromwell's rule and the other at the crucial year of 1700 (a leap year under the Julian rule, but not the Gregorian) also failed.

The act of Parliament that eventually brought Britain onto the Gregorian system in Hogarth's day was crafted by Philip Stanhope, the Earl of Chesterfield, primarily for reasons of convenience: while serving as an ambassador abroad, he was annoyed by the need to convert between Julian and Gregorian dates. He resolved that when he got back to England, he would fix the state of the calendar. Stanhope and his allies skillfully managed the legislative process by maximizing the prominence of astronomical arguments and minimizing the problematic papal connection: the act officially changes the computation of Easter for the Church of England to a system identical to the Gregorian reform, phrased in a way that omits all mention of the original source. The act passed smoothly through Parliament, and Wednesday, the second of September, 1752, was followed by Thursday, the fourteenth, in Britain and its colonies around the world.

Britain wasn't the last country to make the switch; Russia held out into the twentieth century, only making the change after the revolution of 1917, by which time they needed to drop 13 days.† The question of when to celebrate Easter has not been completely settled, either, as many Eastern Orthodox denominations continue to use the Julian rules when computing the date of

---

* Dee justified his system on the basis of astronomy, and also favored dropping an extra day to shift the equinoxes to the dates they had during the life of Jesus, rather than following the Council of Nicaea.

† Both 1800 and 1900 were leap years in the Julian system but not the Gregorian.

Easter. The Gregorian calendar is, however, the most successful of the luni-solar calendar systems ever devised for keeping in synch with the tropical year, the culmination of thousands of years of timekeeping effort. The mere 26 seconds of difference between the length of the Gregorian year (on average) and the tropical year means it will take 3,323 years before the date of the March equinox slips by a full day.

As a matter of formal rules, the transition from the Julian to Gregorian calendars is a tiny change—shortening the average year by 10 minutes by dropping three days every 400 years. However, its development and implementation demonstrate two of the most important points regarding the history of timekeeping: first, the tiny size of the change in the length of the year demonstrates the power of keeping records over a long time—the difference between the Julian calendar and the tropical year was only detectable because of observations made over a span of hundreds of years. The other critical point made by the Gregorian reform is the essentially social and political nature of time. The shift from Julian to Gregorian is tiny, and of little practical significance to things like agriculture. As the Egyptian calendar demonstrates, stable civilizations can last a very long time with much larger errors in their calculation of the length of a year.

The change to the rules, and the political wrangling needed to implement it, was deemed necessary primarily for reasons of theology: a desire to ensure that a particular religious festival would be celebrated at the right time in the right season. This demonstrates the essential arbitrariness of calendar systems and the way they reflect social priorities as much or more than they track astronomical realities in an undeniable way.

As small as the Gregorian reform was, it is not the most dramatic example of a different calendrical system reflecting different priorities. For that, we need to cross the Atlantic for a close look at a radically different sort of timekeeping system, based on a completely different kind of cycle.

Chapter 4

THE APOCALYPSE THAT WASN'T

During the first decade of the twenty-first century, a surprising amount of cultural energy accumulated around a prediction that on December 21, 2012, the world would end, or at least be transformed into something unrecognizable. The exact mechanism by which this was supposed to happen varied from source to source. Some pushed barely plausible natural disasters (giant sunspots, a reversal of the earth's magnetic field, the eruption of the Yellowstone supervolcano);* others uttered completely preposterous conjectures with minimal connection to modern science (the planet being sucked into the black hole at the center of the Milky Way or the solar system entering an "energetically hostile" region of the galaxy). And, inevitably, some said the end times would be caused by aliens.†

The phenomenon was fueled in large part by the burgeoning internet, with predictions proliferating on a dizzying number of websites (most of which are now defunct), but it expanded into more traditional media. Books with titles like *Apocalypse 2012* and *Beyond 2012: Catastrophe or Ecstasy* promoted various theories of what was to come, while television shows on the History and Discovery channels did episodes inspired by the phenomenon. It even provided the hook for the Hollywood blockbuster *2012*, directed by Roland Emmerich

---

* All of these are genuine existential threats to humanity, but none are plausibly predictable.
† Somebody always suggests aliens. It's never aliens.

and starring John Cusack, which made over three-quarters of a billion dollars despite fairly dismal reviews.

The 2012 phenomenon for the most part fits comfortably within a long tradition of eschatology, in which would-be prophets forecast the end of the world by one means or another. What sets the failed apocalypse of 2012 apart from the others, and makes it relevant to this book, is the basis given for the date: where most modern apocalyptic claims hinge on strained interpretations of the Christian Bible, the would-be prophets of 2012 determined the date for their predicted cataclysm by a very straightforward extrapolation from the Mayan calendar.

The classical Mayan civilization recorded dates using a complicated system of multiple cycles, and running this calendar forward, the longest of these cycles would be complete on December 21, 2012. It's this event, the end of the longest cycle recorded by the Maya during the first millennium CE, that various would-be prophets (starting with José Argüelles in 1987) seized on as the date for the end of the world as we know it. From the perspective of 2021, when I'm typing these words, the 2012 episode has a faintly comic air. The world did not experience a notable cataclysm in December 2012, and it was always more than a little farcical to think that the Maya around 600 CE would've been able to foretell disaster for us 14 centuries in their future while remaining unaware of the impending collapse of their own civilization.

The 2012 episode reflects poorly on those who promoted it, but it did have one small bright spot: it directed attention toward the remarkable accomplishments of the Mayan civilization. While their calendar system was not, in fact, useful as a countdown to doomsday, they had a sophisticated system of mathematics, and they were first-rate astronomers. The Mayan calendar and Mayan astronomy were impressive achievements, worth discussing in their own right. The Mayan calendar may also be the best illustration of the social dimension of time.

TIME AS A SOCIAL CONSTRUCT (AGAIN)

Most of the major civilizations of Eurasia—the various empires bordering on the Mediterranean, in Central Asia, and ancient China and India—have used some form of luni-solar calendar. Given that all these cultures had at least intermittent contact with each other, this is perhaps not surprising, but we also find similar ideas at work in very different contexts: in the southwestern United States, for example, the traditional calendar of the Diné relies on lunar months checked by the heliacal rising of certain stars in the winter months, while the people of the Trobriand Islands near New Guinea use a system of lunar months synchronized to the spawning of a particular marine worm during the nights after a full moon in spring. Intercalating lunar months to keep in synch with the seasons is an idea with global reach.

The ubiquity of luni-solar calendars, combined with the impressive precision of the Gregorian calendar, may create the impression that such a system is inevitable. While luni-solar calendars corrected against seasonal drift are a common choice, they are not the only possible approach. In order to really see alternative calendrical systems, we must look into cultures that were free to develop their own approaches to tracking the passage of time. Unfortunately, many of these traditional cultures have been ravaged by decades of European imperialism, with their original practices virtually eliminated before Western academics began to make serious efforts to record and understand them in good faith.

The best place to see the range of different choices made by humans deciding what to prioritize when keeping track of time is in records made before those cultures came in contact with Western empires. Sadly, such records are few and fragmentary. The classic Maya, one of the few Mesoamerican civilizations to leave behind written documents, provide a spectacular example, both in terms of the difference between their priorities and our own, and their success at meeting their chosen goals.

A Brief History of the Maya

"Maya" is the collective name given to a Mesoamerican civilization of city-states that flourished in the Yucatán peninsula (covering the modern nations of Belize and Guatemala, and parts of El Salvador, Honduras, and Mexico), with its peak in the period from about 200–900 CE. They were not the first major civilization in the region (the Olmec civilization, makers of enigmatic giant stone heads, had its heyday in the first couple of millennia BCE) or the most powerful empire of their day (that honor would likely go to Teotihuacán in central Mexico, which overthrew and replaced the rulers of the Maya city-states of Tikal, Copán, and Quiriguá in the years around 400 CE), but they arguably left behind the greatest legacy in art and architecture. And, most importantly, these included a vast number of hieroglyphic inscriptions and a small number of written books, which provide the most detailed record we have of pre-Columbian culture and science.

The Mayan writing system was still in use when the Spanish arrived in the Yucatán in the 1500s, but it was quickly stamped out, along with many other traditional practices. Bishop Diego de Landa, tasked with converting the Maya to Catholicism, recorded with some satisfaction that at his direction Spanish authorities had rounded up large numbers of Mayan books and "as they contained nothing in which were not to be seen as superstition and lies of the devil, we burned them all."* As a result, the few remaining Mayan codices[†] and numerous carved glyphs were indecipherable for years, until they were finally cracked in the 1960s.[‡]

The surviving inscriptions in stone mostly come from tombs, temples, and "stelae," elaborately carved pillars that the Maya erected at regular intervals, which include accounts of (at the time) recent history. As a result, archaeologists have been able to put together a reasonably vivid chronicle of the rise and fall of classic Maya civilization: the rivalry between the city-states of Tikal and Calakmul in the early

---

* From Diego de Landa's *Relación de las cosas de Yucatán*, translated by William Gates in 1937 and published as *Yucatan Before and After the Conquest*. The book was written after de Landa was recalled to Spain to stand trial for running an Inquisition regarded as too brutal even for the Spanish conquistadores, which is really saying something.

† Only four codices remain: three were removed to Europe at some time around the Spanish conquest, the fourth was found in a cave in the 1960s.

‡ Ironically, de Landa played a key role in the decipherment: in his account of his activities in the Yucatán, he included an attempt to correlate Mayan script with the letters of the Latin alphabet. This contained contradictions, and was long dismissed as nonsense, but de Landa had failed to realize that the Mayan glyphs were phonetic, not alphabetic. That realization was crucial to decoding the written language.

period, followed by a later shift of power northward to Chichén Itzá and Uxmal. The final collapse of the civilization remains enigmatic, however, as it's recorded only by the cessation of construction of the stone monuments that provide our best record of Maya activities.

THE MAYAN CALENDARS, PLURAL

In assembling a Maya chronology, archaeologists are aided immeasurably by the clear and consistent dating of many monuments using a system known as the "Long Count." The presence of Long Count dates is one of the defining characteristics of the classic period of Maya archaeology, and it reflects a preoccupation with timekeeping that rivals that of the Hebrew and Julian calendars that led to the modern Gregorian calendar. The Mayan calendar is a sophisticated system of interlocking cycles, but with a dramatically different basis both mathematically and astronomically.

Mayan Math and Dating

It's a bit of a misnomer to speak of "*the* Mayan calendar," as the Maya actually used three slightly different systems in parallel. The most familiar of these is an agricultural calendar known as the *haab*, which had 18 named months, each with 20 days. Dates in the *haab* were given as a number and a month name, much like in Eurasian calendars. A *haab* date would be a number-name pair, like 8 *Ch'en* or 15 *Mak*. Why so many months, and why so short? The 20-day months of the *haab* had, for the Maya, a kind of mathematical elegance.

Unlike our system based on powers of 10, the Mayan mathematical system was based on powers of 20. Where we have ten unique symbols for the numbers 0 through 9, they had twenty, for the numbers 0 through 19:

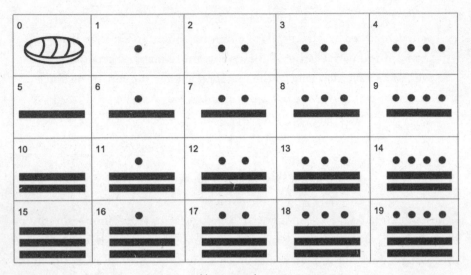

Mayan numbers.

With these symbols, they used a positional arithmetic system very much like ours, but with 20 as the base,* so the smallest digit represented ones, the next multiples of 20, the third multiples of 400 (20 × 20), the fourth multiples of 8,000 (20 × 20 × 20), and so on. In Mayan notation, the number 10,000 would be written:

Dot Bar Shell Shell
That is: (1 × 8,000) + (5 × 400) + (0 × 20) + 0 = 10,000

The 20-day months of the *haab*, then, represent one full counting unit for the Maya, which has a certain appeal. As this was an agricultural calendar attempting to match the tropical year, it needed 18 such months to get to 360 days. Like the ancient Egyptians, the Maya were aware that this comes up well short of a full year, so they, too, included an additional five-day period outside the named months. For the Maya, these *Wayeb* days were deemed particularly inauspicious.

Operating in parallel with the *haab*, they had a second calendar system, the *tzolkin*, based on a set of 20 named days, each associated with a deity that

---

* Base-20 arithmetic shows up in many cultures, probably because 20 is the total number of fingers and toes for a typical human. We even see vestiges of base-20 counting in French, where the word for 80 is literally "four twenties."

"ruled" that day. These cycled through one after another, also accompanied by a number, but in this case, the numbers only range up to 13 before starting over. This combination of number and name was used for divinatory purposes, as seen in almanacs dating from the colonial period.

3	Ben	Bad for those who go through the forest
4	Ix	The queen bee is fertilized
5	Men	Bad
6	Cib	Bad for walking in the forest
7	Caban	Bad. The deer's cry is imitated
8	Eznad	Bad for the people of worship
9	Cauac	Good for bees
10	Ahau	Good. The burner begins the fire
11	Imix	Bad for leaders
12	Ik	Bad. Good wind
13	Akbal	Bad for those who keep watch
1	Kan	Bad. Calm
2	Chicchan	Bad
3	Cimi	Bad
4	Manik	Good
5	Lamat	Good
6	Muluc	The passage of the sun is measured
7	Oc	Bad
8	Chuen	Bad
9	Eb	Good
10	Ben	Bad

In addition to showing the operation of the *tzolkin* list of numbers and names, this list gives a good sense of the cheery outlook of colonial-era Maya . . . From Anthony Aveni's *Empires of Time: Calendars, Clocks, and Cultures* (Basic Books, 1989).

These *tzolkin* dates could be combined with *haab* dates to give a four-character designation to any given day: one significant date recorded at Copan, for example, is given as 10 *Ahau* 8 *Ch'en*. Any such combination of numbers and names would repeat once every 18,980 days—after 52 *haab* years and 73 *tzolkin* cycles—defining a "Calendar Round." This is roughly comparable to a human life span, so any individual person would reasonably expect to see a given calendar round date repeat only once.

For tracking time on the scale of city-states, the Maya used the Long Count, which functioned a lot like the *ab urbe condita* dating of the Romans, or the modern Common Era system, just with more units. For the Long Count, the passage of time was reported in *kin* (the Mayan word for day); *uinal* (1 *uinal* = 20 *kin*, so a *haab* month); *tun* (1 *tun* = 18 *uinal*, a *haab* year less the five unlucky days); *katun* (1 *katun* = 20 *tun*, about two decades); and *baktun* (1 *baktun* = 20 *katun*, or 144,000 days, or roughly 394 tropical years). This is essentially a base-20 progression, with the *tun* shortened to only 18 *uinal* so as to correspond more closely to a tropical year.

Long Count dates are one of the hallmarks of classic-period Mayan art and architecture, and these cycles played an important role in civic life. Many of the stone stelae whose inscriptions form the basis for the modern chronology of Mayan civilization were erected to commemorate the completion of a *katun*; we can think of them as more enduring versions of the "best-of-the-decade" lists generated by news and entertainment outlets every time the Gregorian date reaches a new multiple of 10.

The Calendar Round and Long Count systems operated in parallel, with many dates recorded in both systems. The confirmed Long Count dates are mostly in the eighth and ninth *baktun*, and correlating these with records of astronomical events gives a very precise date for the start of the Long Count according to the Maya: the Mayan date 13.0.0.0.0, 4 *Ahau* 8 *Cumku* would correspond to the Gregorian date August 11, 3114 BCE. The start date of 3114 BCE falls more than 3,000 years before the earliest stelae bearing Long Count dates, making it seem somewhat dubious that this system was in continuous operation for all that time. A more likely possibility is that the Long Count calendar was implemented near the start of the classic period, and backdated to enhance the prestige of the ruling authorities at that time.

This combination of cycles strikes many people accustomed to the Gregorian system as inordinately complex, but it's not without its parallels in Western culture. Perhaps the best modern analogue to the multiple-cycle dating of the Maya is the Julian day system, which despite the name is contemporaneous with the Gregorian calendar. This system was established in 1583 by the French religious leader and scholar Joseph Scaliger as a combination of three regularly tracked cycles (the 19-year Metonic cycle used to determine the intercalation of months in the Jewish and ecclesiastical calendars; the 28-year cycle over which the Julian calendar repeats;* and the 15-year "indiction" cycle associated with military pay in the Roman empire, which continued to be used for civil dating long past the fall of the empire). Many historical events were given dates in terms of the number in each of these (year 9 of the Metonic cycle, year 3 of the indiction cycle, and year 5 of the Julian calendar sequence), allowing the combination to uniquely identify a particular day over a cycle of 7,980 years.

Scaliger "ran back" all these cycles to the last time each was at year 1 at the same time, and his system measures dates by essentially counting the elapsed days since the start of this "Julian epoch." The notional start date he arrived at is January 1, 4713 BCE (if we were to extend the Julian calendar backward; using the Gregorian system it would be November 24, 4714 BCE). This significantly predates even the Neolithic monuments discussed in Chapter 1, but it does not imply that Julian dating was in use for all that time. As surmised above, the Long Count dates of the Maya may similarly have been established by running some of their cycles backward to a particularly auspicious date, creating an impression of a great depth of history for civil and religious institutions of more recent vintage. Archaeoastronomers have proposed a fair number of later dates at which a backdated Long Count might've been implemented, based on particular fortuitous combinations of Calendar Round dates and astronomical events, but there's no conclusive evidence.

Like the Long Count, the Julian day is a straightforward count of elapsed time: for example, January 1, 2000, would be Julian day 2,451,545. As such, it's used "behind the scenes" in some computer systems, with Gregorian dates calculated from the continuously incremented Julian day as needed. The system is also used for recording astronomical events, where the time elapsed between events is of greater importance than the information about the relative position within the tropical year provided by the Gregorian month and day.

---

* The 28-year cycle is the product of seven days of the week and the four-year cycle of leap years. It takes 28 years to cycle through all the possible Julian calendars: seven leap years, one starting on each day of the week.

TROPICAL ASTRONOMY AND THE *TZOLKIN*

Most of the elements of the Mayan calendar have analogues within calendar systems employed in the Western world: the *haab* tracks the tropical year, the Long Count is analogous to the Julian day, etc. The truly unique feature of the Mayan system is the 260-day *tzolkin* cycle, which has no clear origin in the sorts of events Western calendars are designed to track. As a result, numerous archaeologists and archaeoastronomers have speculated on the origins of this cycle.

The most basic explanation is simple numerology: 260 days is 20 (the base number for Mayan mathematics) multiplied by 13, a number with some significance in their cosmology. This has a bit of a chicken-and-egg problem, though, in that it's not clear whether the *tzolkin* cycle includes 13 numbered days because 13 had a preexisting mystical meaning, or whether the number 13 acquired importance because of its use in the *tzolkin*.

Another possibility is that the *tzolkin* is derived from the tracking of some other interval. The number 13, for example, could be based on a count of lunar months, each given 20 days in an attempt to harmonize with base-20 arithmetic. Or the origin might be biological: the 260-day length of the *tzolkin* is not too far off the length of a typical human pregnancy, so it might have its origin as a tool for tracking gestational age.

For the purposes of this book, however, the most intriguing possibility is an astronomical origin: the notion that the *tzolkin* cycle could be based on the motion of lights in the sky. And, in fact, there is a good candidate for such an astronomical origin, arising from the unique features of life in the tropics.

For a naked-eye astronomer in the middle latitudes, like most of the cultures we have discussed to this point, the seasonal motion of the sun is defined by four events: the June and December solstices and the March and September equinoxes. At other positions on the globe, though, additional phenomena appear and can dominate the experience of the seasons. The most spectacular example of this is that for any latitude above about 66.5 degrees (the location of the Arctic and Antarctic circles) there is at least one winter day when the sun does not rise, and at least one summer day when it does not set.

Closer to the equator, between 23.5 degrees north and south latitude, the new phenomenon is less extreme but still readily apparent to a careful observer. In the tropical latitudes, there is at least one day a year when the sun at noon is directly overhead. Exactly on the tropic, this occurs only on the day of the summer solstice, as in the story of Eratosthenes's measurement of the earth's circumference (back in Chapter 1). Between the tropics, the noonday sun is sometimes north of the zenith and sometimes south, depending on the season, so there will be two "zenith-crossing" days when the sun is directly overhead at noon.

In the region north of the equator and south of the Tropic of Cancer, the north-south motion of the rising and setting positions of the sun follows the same pattern as at higher latitudes, though it covers a smaller range. The daily pattern of the sun's motion is also the same in Central America as farther north: the rising sun moves up and to the right, shifting south as it climbs the sky, only not as far south as it would move at higher latitudes.

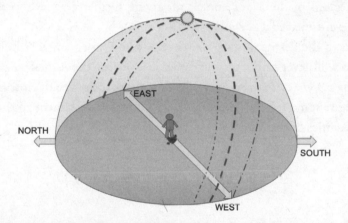

Motion of the sun in the tropical latitudes. For one day between the equinox and the solstice, the sun rises north of east, is directly overhead at noon, and sets north of west.

On the two zenith-crossing days, the sun rises north of east and moves south by just enough that at its highest point it is directly overhead and vertical objects cast no shadow at noon. At a latitude of about 14.5 degrees north, in the region where the Maya civilization arose and flourished, these occur in April and July and are separated by 105 days; the start of this period roughly corresponds with the start of the rainy season, a key date for agriculture in the Maya heartland. For the other 260 days of the year, the sun spends the entire day south of the zenith. This 260-day period, when crops were harvested and most warfare conducted, could be a plausible origin for the *tzolkin*, as suggested by the archaeoastronomer Anthony Aveni (among others).

There is evidence, both archaeological and testimonial, that the zenith crossing date held some significance for the Maya. Colonial-era sources report that the celebration starting a new year of the *haab* fell in July, near the second zenith crossing.* There are also "zenith tubes," vertical shafts allowing sunlight to reach an otherwise dark room on dates near the zenith crossing, found in Mayan structures at Monte Albán and Xochicalco in Mexico, and numerous buildings are aligned to face the position of sunrise or sunset on the dates of the zenith crossing. One of the handful of slit windows in the famous Caracol observatory at Chichén Itzá, for example, faces toward the sunset on zenith passage dates. The nominal start date of the Long Count calendar, on August 11, 3114 BCE, also falls quite close to the second zenith crossing date in the southern regions of the Maya homeland, giving further credence to the idea that these events played a key role.

Thanks to the destruction of Mayan records by time, conquistadores, and looters, we will never know the definitive reasons for the implementation of the 260-day *tzolkin* cycle. It seems likely, though, that the zenith crossing dates played at least some role, making the *tzolkin* in part a testament to the astronomical activities of the Maya.

SOME OTHER BEGINNING'S END

In May of 2012, just seven months before the end of the thirteenth *baktun* of the Mayan Long Count, a team of archaeologists announced the discovery of what may have been a workroom for astronomical priests and scribes in the Maya city of Xultún, in northern Guatemala. The structure dates from about 800 CE and includes columns of numbers painted on the plaster walls that are reminiscent of the astronomical tables of the Dresden Codex (one of the handful of surviving Mayan books, which somehow escaped the colonial-era purges and ended up in Germany—more about these in Chapter 7), but predate them by several centuries. Analysis of the walls suggests that they were replastered

---

* Given the drift of the 365-day *haab* relative to the tropical year, this may have been a coincidence.

multiple times to allow new calculations, like a very slow version of the white-boards that adorn the office walls of many modern astronomers.

The Xultún murals are badly degraded* and seem likely to have been a work in progress, so it's difficult to interpret exactly what they mean. The calculations include a set of *tzolkin* dates followed by intervals (in Long Count format) that are round multiples of the main cycles tracked by the Maya: the *haab*, *tzolkin*, and Calendar Round; the cycles of Venus and Mars; and the cycles of lunar eclipses. These would've been useful to astronomer-priests looking to calculate future astronomical omens in a way that harmonizes with the 260-day *tzolkin* cycle. The intervals recorded also cover spans of time far greater than that from the time of their writing to the present. The longest is more than 6,700 years, greater than the full span from the nominal Long Count start date in 3114 BCE to the supposed end of the world in 2012. These strongly suggest that the Maya did *not* consider December 21, 2012, to be the date of some world-ending cataclysm.

The Xultún astronomical calculations were painted on the wall late in the classic Maya era, at a time when some of the city-states that had dominated the period were well into their final collapse. The enormous intervals involved as well as the conditions at the time of their writing show the importance of projecting their cycles of time into the distant future, offering a hint at the worldview of the people behind the calendar system, one very different from our own. To the Maya, the Long Count was not a doomsday countdown to a singular catastrophe, but an infinitely extensible set of repeating cycles. In the somewhat poetic phrasing of William Saturno of Boston University, who led the Xultún expedition, people in Western culture "keep looking for endings, [while] the Maya were looking for a guarantee that nothing would change. It's an entirely different mindset."[†]

---

* It's a minor miracle that they're preserved at all; they appear to have survived only because the room was deliberately buried in a later renovation of the temple complex and remained underground until partially exposed by looters.

[†] The quote appears in many news stories about the find; see, for example, the BBC's May 10, 2012, write-up: https://www.bbc.com/news/science-environment-18018343.

This starkly different approach to time, together with the unique 260-day basis for the *tzolkin* calendar,* stands as a particularly dramatic example of the social dimension of time: that what we choose to emphasize in our calendrical systems is a product as much of the cultural context in which we operate as any objective facts about seasons and planetary orbits. To modern eyes accustomed to the Gregorian calendar, the Maya seem inordinately fond of the numbers 13 and 20, but were our positions reversed, it's likely they would be bemused by our disconnection from the essential natural rhythms that repeat on a 260-day cycle.

The tables in the Dresden Codex also provide an essential reminder that astronomy, and science in general, is not a unique invention of Western civilization—every civilization we have knowledge of has developed their own collection of astronomical knowledge and associated methods of tracking time. This universality of science also extends to more earthbound activities; in the next chapter, we'll step back from astronomy for a bit to consider another essential set of timekeeping technologies, which reached their highest level of development in a place far removed from modern Europe.

---

* This is shared with and was possibly passed on to other Mesoamerican cultures. The Maya writing system and practice of erecting engraved stelae at regular intervals means that their system is by far the best documented.

Chapter 5

DRIPS AND DROPS

The astronomical cycles we have discussed thus far unfold in a stately and silent way, across days and months and years. For people in modern societies, though, solstice markers and calendars are not the first objects we picture when the subject of timekeeping comes up. When we think of timekeeping technology, for the most part, we envision something that is both faster and showier. What immediately comes to mind for most people when the subject turns to timekeeping technology is a clock: a device for measuring time on much shorter intervals than a day. We use clocks for numerous purposes: to limit work hours or sporting competitions, to coordinate meetings with friends and colleagues, or to track the unfolding of rapidly changing events, and so on. Clockmakers employ a wide variety of displays to achieve this end, from lighted displays to ringing bells to musical alarms.

Above all, though, the time measured by clocks is a *public* resource. Astronomical timekeeping requires expert knowledge to identify stars and their positions or to work through the calculations in the Dresden Codex to determine the next appearance of Venus. In many societies, this was carefully guarded knowledge, closely held by elites and used to enhance their power. Clocks, on the other hand, require minimal skill to read, and the time they show is coordinated worldwide so that everybody knows what time it is.* Many clocks

---

* We'll discuss how this coordination is accomplished in more detail in Chapters 11 and 13.

are landmarks, built into churches and civic buildings, with large public faces to display the time in easily readable form to the world at large, and ringing chimes to announce the hour even to those without a clear view.

The public clocks that govern so much of modern life are mostly either mechanical or electronic, but the need for open public timekeeping stretches back thousands of years before the invention of these technologies. As a result, technology for measuring shorter intervals in a reliable way has been developing in parallel with astronomical and calendrical science for several thousand years. The central concept behind these clocks is surprisingly simple, remarkably ancient, and nearly universal.

HOW TO KEEP TIME IN THE DARK

A well-constructed sundial can, when aligned and used properly, give the time of day to within a few minutes, and as such sundials were used as the ultimate time reference across many cultures over a span of centuries. In large cities, the civil time was announced to the general population at regular intervals via loud noises: blowing trumpets or beating drums to signal the change of hours. These hours were regularly calibrated against solar time.

The use of the sun to mark time, though, has one glaring flaw: it requires the sun to be shining. On a cloudy day, the sun doesn't cast the clear shadow needed to read a sundial with any accuracy, and at night or indoors there are no shadows. As we've seen, the motion of the stars can be used to track time at night, but as the visible stars change with the seasons, this requires a good deal of observational knowledge and training (and, of course, all that training is useless on a cloudy night).

Some backup method is needed, then, for tracking time when the sky can't be used as a timekeeper. Cultures all around the world have confronted this problem and mostly come up with the same solution: filling a container with some amount of stuff and letting it run out. Time intervals can be measured by the amount of stuff that's run out, and as long as the container, the material, and the outlet are reasonably standardized, this can provide a reliable "tick" to track time over intervals ranging from minutes to hours.

This is the core operating principle of the water clock. Water clocks of various types served as state-of-the-art timekeepers for thousands of years and were the subject of an enormous amount of engineering innovation through that time. Specially engineered water clocks were used for the full range of timing applications we recognize today: tracking the slow beat of civil time through dark nights and cloudy days, ringing alarms to announce the time or wake the sleeping, and acting as stopwatches to limit the time allotted to activities or serve as precision measurement devices in scientific experiments. All of these variations, across several millennia, have their roots in the same simple idea: filling a leaky container with water and tracking time by how much leaks out. The "tick" of a water clock comes from a single core process, the flow of liquid out of some container, but there are two different methods for providing the necessary readout. An outflow water clock tracks time by the dropping water level in the reservoir as it empties, while an inflow water clock records the filling of a second container positioned to catch the water leaving the first.

The oldest surviving water clock is of the outflow type, an alabaster vessel shaped something like a modern flowerpot, found at the Karnak temple in Egypt. Decorations on the outside of the vessel include the name of the Pharaoh Amenhotep III, dating it to about 1350 BCE. The interior of the Karnak "clepsydra"* is inscribed with hourly markings, suggesting that this was intended to mark time at night to ensure that temple rituals were performed at the appropriate moments. While the Karnak clepsydra is by far the best-preserved example of an ancient water clock, it's by no means unique. Fragments of similar devices have been identified at a number of sites around the Mediterranean, with dates of origin spanning some 2,000 years, well into the Roman era. There's even some hint as to the origin of these devices in an inscription found in the tomb of an Egyptian official named Amenemhet during the reign of Amenhotep I, around 1500 BCE. Listed among the accomplishments of Amenemhet's life is a claim that he had invented a water-based timekeeping device to track the changing length of the night through the course of the year. While there are few details given, the Karnak clepsydra matches this general description.

---

* The word "clepsydra" is from the Greek for "water thief," and it was the term for outflow-type water clocks around the ancient Mediterranean.

More commonly in the ancient world, outflow-type water clocks served the same purpose as a modern stopwatch. Law courts in Athens and Rome made use of outflow-type water clocks to limit the speeches of advocates: a set amount of water would be added, and when the flow of water stopped, so did the argument. Court records from Athens include occasional remarks where speakers comment that they see by looking at the clock that their time is nearly out. Since the surviving clocks we have were opaque earthenware, it's not likely they were able to see the water level inside; rather, they could judge the level by the strength of the water stream, which brings us to one of the crucial problems of outflow-type water clocks. In a typical clepsydra draining from a hole near the bottom, the rate of the water flow depends on the depth of the water. This is easily illustrated by, for example, poking a hole in the side of a large paper cup and setting it on the edge of a sink to drain. When the cup is nearly full, the water jets out in a stream that can extend well into the sink basin, but as the level drops, the flow slows, until the last few drops merely dribble out. The last few milliliters of liquid take far longer to drain out than an equivalent amount when the cup was nearly full. Thus, an outflow-type clepsydra is a clock that ticks more slowly as it goes.

One technique for dealing with this change in rate is to use an inflow-type clepsydra instead, tracking the time needed to fill a container of standard size from a reservoir that's large enough that the draining doesn't significantly decrease the level. Inflow-type water clocks were developed in both Ptolemaic Egypt, around 300 BCE, and in China, around 200 BCE, and over the next millennium and a half, clockmakers developed numerous refinements to allow these to serve as precision timepieces. Inflow-type water clocks were the state of the timekeeping art for nearly 2,000 years, until they were supplanted by mechanical clocks.

In fact, one variant of the inflow-type water clock continued to see use well into the twentieth century. This was the Persian *fenjaan*, consisting of a metal bowl with a small hole in the bottom that was floated atop a reservoir of water. As the bowl slowly fills with the liquid, it dips lower in the water and then suddenly sinks. These sinking-bowl clocks were traditionally used to regulate the distribution of water from communal irrigation systems in rural Iran. Water flowing from a shared source would be directed to a particular farm, the fenjaan would be placed in a reservoir, and respected local officials were tasked to

ANCIENT ENGINEERING TO MODERN PHYSICS

If the goal is to make a stopwatch or countdown timer that repeatably measures a particular time interval, an outflow-type water clock is an excellent solution, which is why these were in common use for millennia. But if the goal is precision timing of an arbitrary duration or of shorter periods within a longer interval, the changing flow rate is a major obstacle. If you fill an outflow clepsydra with half the original amount of liquid, the time needed for it to run out is not half the original time, but considerably longer than that.

In modern physics, we understand this as a consequence of Pascal's law in fluid mechanics, which says that pressure (the force on an object divided by the area over which the force is applied) applied to an incompressible liquid is transmitted uniformly throughout the liquid. If you exert a one-newton force* on the top surface of a one-meter cube of water (thus applying a pressure of one newton per square meter), that force is transmitted equally to each of the other surfaces. The pressure on each of the walls increases by one newton per square meter, so the force pressing down on the bottom of the cube increases by one newton, as does the force pushing out on each of the sides.†

In the case of an outflow-type water clock, the pressure that matters is not a push being deliberately applied, but simply the weight of the water itself. We can see this by thinking about a simplified water clock consisting of a tall tube with straight sides and an exit nozzle pointing horizontally out. To determine the behavior of the water exiting this "clock," we can look at just the tiny bit of water right at the mouth of the nozzle and ask what the pressure acting on it is. The area will be the area of the nozzle, and the force will be the combined weight

---

* One newton is the force required to accelerate one kilogram at a rate of one meter per second per second. It's also, if you want to be cute, roughly equal to the gravitational force on an apple with a mass of 100 grams (3.5 ounces).

† This is the principle behind a hydraulic jack, which consists of a thin liquid-filled tube connected to a large liquid-filled piston. When a force is applied to the tiny area of the tube, the resulting pressure is transmitted through the liquid to the surface of the piston. A one-newton force applied to liquid in a tube a bit more than a millimeter across leads to a pressure of one newton per square millimeter across the piston head, for *every* square millimeter in the head. A piston head 11 centimeters across has 9,500 times the area of the tube, so such a system would supply a 9,500-newton force easily large enough to lift a car.

of all the water above that point.* The taller the water column, the greater the weight, and thus the greater the pressure forcing water out the exit nozzle.

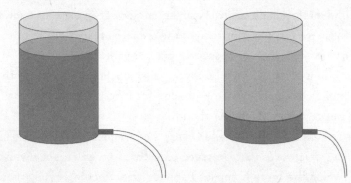

In a mostly-full container draining through a nozzle near the bottom, the weight of the water above the nozzle produces a high pressure, causing the water to jet out a significant distance horizontally. When the liquid level is lower, so is the pressure, and the output falls more vertically.

The exact rate of flow will depend on the size, shape, and composition of the nozzle itself in a complicated way. For a given configuration of those things, though, the changing pressure as the container empties will determine the change in the rate of flow, with the velocity of the exiting water changing like the square root of the height of the water column. Thus, a clock filled to half the original depth will see water leaving the nozzle at 70 percent of the original rate, and require a longer time to drain a given amount of liquid.

This changing rate of flow can be partially compensated for by changing the shape of the container used to make an outflow-type clock. We can see this in the Karnak clepsydra, whose hours are marked at roughly equal height intervals on the inside of a container whose walls slope outward. Early on, when

---

* Strictly speaking, we also need to include the effect of atmospheric pressure pushing down on the top of the column, and inward on the water in the exit nozzle, but when the top is uncovered these are equal and they cancel out. Atmospheric pressure becomes important, though, when one end is open to the air and the other is not, which is why you can pick up liquid in a drinking straw by covering the top with a finger. Your finger blocks the pressure at the top of the straw, but not at the bottom, so as long as the weight of the water inside is not large enough to overcome the 100,000 newtons-per-square-meter pressure pushing up on the bottom of the straw, the liquid remains trapped inside.

the clepsydra is nearly full, a larger amount of water needs to flow out to cause the fill level to drop by a given distance compared with later on, when the total fill level is lower and the container is narrower.

Calculating the ideal shape to ensure an equal flow rate in an outflow water clock is a relatively simple problem today thanks to the invention of calculus, but that was still more than 3,000 years off when the Karnak clepsydra was built. The actual shape of the Karnak clepsydra was probably found by empirical means, but it's not a bad match to the theoretical ideal across the region where the hour marks occur. Experiments with models made from plaster casts of the original suggest that the Karnak clepsydra worked very well at its (presumed) function of dividing the night into equal hours. As an additional refinement, the Karnak clepsydra features not just one hourly scale, but 12, each headed with the name of a month in the Egyptian civil calendar. The spacing between hour marks on these scales varies in a way that tracks the changing length of the night across the seasons.

The careful shaping of the Karnak clepsydra is a significant accomplishment for the scientists and engineers of 1500 BCE, fully justifying Amenemhet's post-humous pride in his invention. There's a simpler way to ensure a constant rate of water flow, though, namely by keeping the level of the liquid constant. Of course, this can't be used in an outflow-type clock (by definition), so it requires a transition to an inflow-type water clock.

The simplest sort of inflow-type water clock uses a very large reservoir of water, so that while the level of the source drops over time, it doesn't change by enough to significantly change the flow rate. Some argue that mentions of water clocks in Babylonian clay tablets refer to a clock of this type, but the archaeological evidence is ambiguous. Most of the ways to ensure a constant water level involve placing an intermediate container between the water source and the container whose filling provides the readout of the inflow-type clock. The simplest type uses a container with a slow drain at the bottom and an over-flow outlet at the top, fed from a source that supplies water at least as fast as it flows out to the readout vessel. The overflow outlet ensures that the water level, and thus the rate of flow into the readout vessel, remains constant as long as the water continues flowing.

The overflow method can be wasteful, unless the spilled water is collected and returned to the source, so various means were developed to maintain a constant water level with less spillage. A simple and elegant approach, developed only in China, is to have a long series of identical containers arranged like a staircase, with each draining into the next. As long as the first container is periodically refilled (say by returning the water collected in the inflow vessel), a container four to five steps down the line will maintain a constant fill level to an excellent approximation, and can be used as a constant-rate source for an inflow clock.

The most famous solutions to the problem of maintaining a constant rate of flow are more complicated in ways that foreshadow the future development of mechanical clocks. Maybe the best-known of these is the floating valve design attributed to the engineer Ctesibius, who lived in Alexandria around 250 BCE.* In this scheme, the constant water level in the intermediate tank is maintained by a float supporting a rod that plugs the outlet of a larger reservoir. When the level in the intermediate tank drops, the plug moves down and water is allowed to flow in, pushing the float back up until the plug cuts off the flow again.

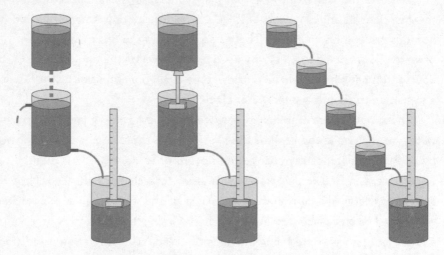

Constant-level inflow water clocks. Left: overflow method; Center: floating plug à la Ctesibius; Right: multi-container cascade from China.

---

* Ctesibius's work was cited admiringly by the Roman engineer Vitruvius and was used in state-of-the-art water clocks for many centuries.

The floating valve employed by Ctesibius points to a key innovation in the readout of inflow-type water clocks. Over a period of centuries, in cultures across Eurasia, makers of water clocks began to develop more and more elaborate ways of indicating the elapsed time, based mostly on floats placed in the readout vessel. These began as simple arrows pointing to a vertical ruled scale and evolved to include gears turned by a rising float, turning a rotating pointer (the forerunner of the hands on the face of a modern analog clock).* The clocks also began to include alarm features such as ringing bells when a target fill level was reached, and even the triggering of elaborate automata to announce the hours with choreographed movements.

---

* An interesting variant from about 1000 CE—the work of the Muslim polymath Ibn al-Haytham—is based on a cylinder that slowly sinks as it fills with water via a small hole in the bottom. The sinking cylinder was connected to a dial pointer by a string over a pulley.

watch the bowl and announce its sinking, at which point the water flow would be redirected to the next farm. As a system with no moving parts to wear out or batteries to run down, this provides an elegant and nearly foolproof method of equitably sharing the flow of water, and also serves as a reminder that the simplest solutions are often the most enduring.

THE TOWER CLOCK

While elaborate water clocks were developed all around the Mediterranean, in both Christian and Muslim nations, the pinnacle of water-driven clock engineering was probably reached around 1100 CE in China. This was the famous tower clock of Su Song, an ingenious device combining the best of water-clock engineering with elements that would become essential for the later design of mechanical clocks (as we will see in the next chapter).

Su Song (1020–1101 CE) was a court official in the later years of the Northern Song dynasty. In addition to being a successful bureaucrat and administrator, he was something of a scientific polymath, publishing a sky atlas mapping the positions of more than 1,400 visible stars, and compiling an important treatise on pharmacology.* While copies of both the star map and the pharmacopeia survive today, the work of Su Song's that looms largest in the historical imagination is an astronomical clock that stood for barely more than 30 years in the Northern Song capital of Kaifeng. The physical clock was lost when Kaifeng fell in 1127, but its workings were described in another of Su's books in sufficient detail to enable modern reconstructions of what was truly an ingenious milestone in the history of timekeeping.

Su Song's interest in timekeeping stemmed from an incident in 1077 CE, when he was dispatched to the north to offer official greetings to the ruler of the Liao kingdom on the winter solstice. Su and his message arrived a day early because the calendar used by the Northern Song at that time was in error. This would've been a great embarrassment to the imperial delegation, but thanks to a

---

* Traditional Chinese remedies incorporate a very wide range of components, giving "pharmacology" a very expansive definition that includes botany, zoology, and even metallurgy.

combination of diplomatic skill and astronomical erudition, Su Song was able to
convince the Liao to accept the greetings
and avoided a major diplomatic incident.
Upon his return to Kaifeng and report to
the emperor, the officials responsible for
the calendrical error were punished, and
Su Song set about a program of reform to
ensure that nothing of the sort would happen in the future. In addition to his own
prodigious talents, he recruited the best
astronomers, mathematicians, and engineers the empire had to offer, most notably
one Han Gonglian, and bent their efforts
toward the construction of a mechanical
timekeeper that would allow the Northern
Song court to track the motion of the stars

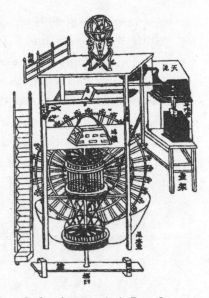

Su Song's tower clock. From Su
Song's *Xin Yi Xiang Fa Yao* (1092).

and display the time and date with unparalleled accuracy.

The final form of the clock was a massive tower, over 12 meters in height,
supporting a giant bronze armillary sphere (an astronomical instrument consisting of a set of concentric rings marked with the angular coordinates of objects
in the sky) on its top platform. In an interior chamber just below the main
platform was a somewhat smaller celestial globe marked with the positions of
the visible stars. The primary purpose of the clock was astronomical, rotating
these devices in time with the motion of the stars, but it also had a public time
display. Large rotating wheels would bring painted figures holding placards displaying the date and time into windows facing out into Kaifeng. Certain hours
were also marked by the sounding of gongs or drums by spring-driven jack
work, making it an early example of a fully automatic civic timekeeper.

The tower clock was essentially an inflow-type water clock, filled from a
constant-level tank system. The massive scale of the astronomical devices—the
armillary sphere alone would have weighed 10 to 20 tons—and the gearing
involved in the time display was far too much to be driven by the float system
used in other water clocks around the Mediterranean, so the tower clock used
the weight of the flowing water to drive the rest of the works.

Its centerpiece was an enormous waterwheel, some 3.35 meters in diameter, bearing 36 scoops around its circumference. When the wheel rotated forward to bring an empty scoop under the flow from the constant-level tank, a system of wooden rods would drop down to lock the wheel in place. When the weight of water in the scoop exceeded that of a counterweight, the scoop would tip downward, triggering a mechanism that would briefly lift the locking rods out of the way, allowing the weight of the filled scoops in the lower quadrant of the wheel to pull it forward,* bringing a new empty scoop under the water, and beginning the cycle again.

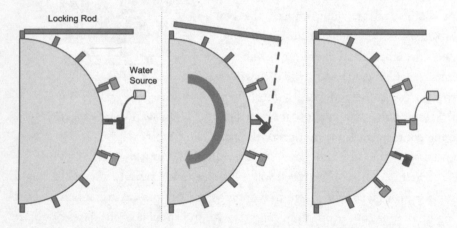

Scoop escapement system from the tower clock. When a scoop fills with water, it tips forward, triggering a mechanism that lifts the locking rod and allows the wheel to turn. When the next scoop rotates into position under the constant-flow water source, the locking rod drops back down and holds the wheel in place until the scoop is full.

This scoop-and-locking rod system is an early and innovative example of an escapement, a system for regulating the driving force powering a clock according to the motion of some component that marks out a regular period. In Western mechanical clocks, the drive and regulation are generally separated: the drive comes from a falling weight, for example, while the regulation comes from

---

* As the scoops tipped and reached the bottom of the wheel, they would empty into a collection tank below, from which water would periodically be returned to the upper reservoir of the constant-level tank system using a hand-cranked scoop wheel.

a pendulum that moves an escapement in and out of the teeth of the clockwork gears. The genius of the Chinese system is that it combines the drive and regulation into a single element: the weight of the water is what pushes the wheel forward, and the filling of the scoops from a constant-level tank ensures that the wheel slips forward at a regular rate (one scoop every 24 seconds, according to modern calculations based on Su Song's text). This makes it, as historian Joseph Needham put it, "a 'missing link' between the time-measuring properties of liquid flow and those of mechanical oscillation."*

Su Song's clock was very much part of a Chinese tradition of wheel-driven timekeepers; Su Song himself referred to a famous clock made by Zhang Sixun around a century earlier (which we'll return to later in this chapter) that also featured a wheel driving an armillary sphere, celestial globe, and jack work to sound the hours. This, in turn, drew on the earlier work of Tang dynasty clockmakers, who were driving celestial spheres with flowing water in the mid-700s CE.

What makes Su Song's clock particularly notable is its massive scale and impressive precision, which resulted from a sophisticated engineering design and careful testing. Su Song and Han Gonglian began with a small wooden model of the clock, then operated a full-scale wooden model for a couple of years, testing its performance against four other types of water clock and astronomical observations before casting the massive bronze parts for the final clock.

The final clock and Su Song's monograph on its workings were complete by 1094 CE, the crowning achievement of a long life in scientific service to the empire; Su Song died in 1101. Sadly, his clock did not long outlive him: in 1126, Kaifeng was besieged and then sacked by the Jin, marking the end of the Northern Song dynasty. Elements of the empire relocated south to a new capital in Hangzhou, but despite the efforts of the best engineers available, among them Su Song's son Su Xi, they were unable to reproduce Su Song's masterwork. The original tower clock was disassembled in 1127 by the conquering Jin and carted north to their capital in Beijing, but their engineers were unable to get

---

* Needham's enormous comprehensive *Science and Civilisation in China* (1954) remains one of the definitive sources on Su Song's clock, and genuinely deserves the overused adjective "magisterial."

  It should be noted, though, that there's little evidence that any clockmaker in Europe was influenced by or even aware of Su Song's magnificent clock.

it working again. The giant armillary sphere and celestial globe remained on display in the city until it fell to the Mongols in 1264.

The tradition of waterwheel-driven clocks in China continued through the Mongol era and into the Ming dynasty, though generally on a less ambitious scale than that of Su Song's tower clock. The Ming era in particular saw one interesting innovation that leads nicely into our next timekeeping system: the replacement of flowing water by flowing sand.

. . . SO ARE THE DAYS OF OUR LIVES

Clepsydrae were the timekeeping workhorses of the ancient world, and the elaborate water-driven clocks produced in the Mediterranean and China were marvels of reliable engineering. There is, however, one major issue that limits the effectiveness of water-based clocks, namely that they require liquid water. While this isn't a huge problem in the warm climate of Egypt, in the winter months in colder climates, water clocks of either type have a tendency to freeze up.

This was a problem well known to Chinese clockmakers: the tabletop clock built by Zhang Sixun in the mid-900s CE (and described by Su Song as one of the influences on his great clock) was built to run on mercury rather than water. Despite being a heavy metal, mercury is a liquid at room temperature, and it remains a liquid down to nearly 40 degrees below zero.* This wide temperature range is why mercury was a traditional working fluid for thermometers, and it allowed Zhang Sixun's clock to operate reliably in even the coldest weather.

While it has some nice features as a clock-making component, mercury is both expensive and highly toxic—the idiom "mad as a hatter" can be traced to the symptoms of mercury poisoning in hatmakers who used mercury compounds to cure felt. A much more economical alternative is to replace the liquid water with a granular substance, such as fine sand. Sand with a sufficiently small grain size will flow through an aperture very much like water, with the very

---

* It doesn't matter much which temperature system you use in this case, as –40° is where Fahrenheit and Celsius scales intersect.

great advantage, from a timekeeping perspective, that it doesn't freeze in cold weather.

Sand has another advantage beyond its relative insensitivity to weather, namely that the flow rate does not depend on pressure. The flow of a granular material like sand is very different than that of a continuous fluid: the individual grains move independently, and they can rearrange themselves to take up less space when pressed on, so pushing on the top of a sand pile does not force the grains at the bottom out any faster. Each grain in a stream of sand draining from a reservoir falls in response to its own weight, more or less, making the rate of flow independent of the depth of the sand pile above the opening.

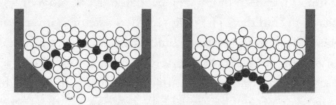

The phenomenon of "jamming," where particles draining through a nozzle can temporarily come together to form an arch that stops the flow.

This is not to say the motion of a single grain of sand is totally unaffected by the presence of the other grains in the flow. Granular flows through small apertures are subject to "jamming," a phenomenon in which individual grains will lock together like puzzle pieces to form a sort of arch spanning the aperture, each pushing on the others in a way that at least temporarily stops the flow. This makes the flow rate dependent on the size and shape of the aperture through which the sand must flow, a problem noted by Chinese clockmakers: sometime in the mid-1500s, an astronomer named Zhou Shuxue was dissatisfied with his sand clock thanks to this jamming behavior, but increasing the size of the opening and changing the gearing to compensate allowed him to make a clock that performed well. Clockmakers in China made the switch from water to sand at the start of the Ming dynasty, around 1370. Save for the change in material, these clocks functioned very much like Su Song's water clock or Zhang Sixun's mercury clock, with the weight of the sand acting as both power source and regulator, turning a wheel to drive the gears that drove the visible display.

In Europe, the primary use for flowing sand was not to power elaborate clocks but simple hourglasses,* where the emptying of one bulb into the other marks the passage of a set amount of time. The exact amount of time required depends on the amount of sand and the width of the orifice between bulbs; as the name suggests, these were typically built to measure times from a few minutes to about an hour. Given the simplicity of the idea, it's surprising that sandglasses were a rather late development in the history of timekeeping, with the first documented appearance of a sandglass in Europe being a 1338 fresco by Ambrogio Lorenzetti titled *Il Buon Governo* in Siena, Italy. The sandglass held by the allegorical figure of Temperance is instantly recognizable, suggesting that these timers had been in use long enough to reach more or less their final form by that time, but there's no clear record to date their invention.[†]

Whenever they were invented, sandglasses became nearly ubiquitous as devices for ordering activities of relatively short duration: timing sermons and other speeches, measuring cooking times in recipes, or timing breaks from work. The most significant drawback of the sandglass as a timekeeper is that for most shapes, it's an all-or-nothing device: sand tends to drain from the center of the pile, making a slowly expanding conical hole. This "plug flow" means that the visible level of sand against the walls of the glass doesn't begin to drop until very late in the process, making it nearly impossible to usefully subdivide the interval of the full glass.[‡] Multiple glasses were sometimes mounted in a single frame to address this problem: a set of four with running times of 15, 30, 45, and 60 minutes, for example, to allow the user to track quarters of an hour.

In the same way that outflow-type water clocks need regular refilling, sandglasses need to be turned regularly if one wants to use them to measure time periods longer than the duration of a single glass. In situations where this was necessary, mostly involving ships at sea tracking time for navigational purposes

---

* European water clocks mostly used float systems to trigger their showier displays, and these don't allow a straightforward replacement of water with sand.

† Some authors attribute the invention of the sandglass to a French monk in the late 700s CE, but there's no real support for this. The utter absence of anything identifiable as a sandglass in the next 500 years of art and literature tends to suggest this story is a legend invented after the fact.

‡ It is possible to make a long and narrow sandglass where the visible level drops in a regular way, but there aren't many historical examples of sandglasses with the correct sort of shape, and the physics needed to predict the shape wasn't developed until much later.

(which we'll discuss more in Chapter 9), pages were delegated to watch the glass and turn it when necessary; generally, arrays of several overlapping glasses would be used as a hedge against lapses of attention that might delay the turning.

Due to their susceptibility to jamming, the filling material for sandglasses was the subject of much experimentation in Europe during the late medieval period: there are records showing the use of powdered marble, silver, tin, and lead, filings of iron and copper, fine emery, and ground cinnamon (which must've made for an incredibly expensive clock), in addition to ordinary river sand. A particularly prized material was "Venetian sand," an extremely dense powder combining charred tin and lead, which flowed much more slowly than other materials (a 24-hour glass of Venetian sand was about the size of a one-hour glass of ordinary river sand). A more economical but still highly regarded alternative was powdered eggshells, which was particularly popular for marine sandglasses.

Sandglasses had a good run as the most reliable and portable timekeeping technology from the 1300s up until around 1600, and a bit longer on board ships; the British navy continued to use sandglasses to calibrate mechanical watches up until 1839. For all their advantages, though, the accuracy of sandglasses was very limited, and they were quickly replaced once European scientists and engineers hit on the idea of purely mechanical clocks, to which we will turn next. There is something powerful and poignant about the operation of a sandglass, though, with the slowly dwindling sand in the upper reservoir serving as a visible reminder of the passage of time. For this reason, while the era of the sandglass as a precision timekeeper has long passed, it continues to enjoy a long afterlife as a metaphor for the passage of time, whether in the credits of a long-running soap opera, or as an icon to indicate that your computer is slowly working through some thoroughly modern problem.

Chapter 6

TICKS AND TOCKS

From sandglasses, we now turn to the other great timekeeping metaphor we've been drawing on for this book: a mechanical clock whose audible "tick-tock" provides us with the term for the repeating process at the core of any timekeeping method. These are still in widespread use both as public landmarks (the clock in Union's Memorial Chapel that I've mentioned keeps time with a swinging pendulum) and in private homes (we have a pendulum-based wall clock in our dining room whose chime we use to limit our kids' activities: "Next time the clock bongs, the tv is going off!"). Our attachment to mechanical clocks is such that many electronically driven clocks will artificially create a ticking sound, though nothing in their actual operation requires it.

In this case, the metaphor extends beyond the clock to encompass the entire world. These days, it's very common for scientists and science writers to speak of the operation of the universe in terms of "clockwork." The solar system, the galaxy, even the universe as a whole is cast as a mechanical clock: an intricate system with innumerable tiny parts moving in complex ways that ticks along smoothly and reliably from the known past into the predictable future. It operates according to simple rules and principles, and its various patterns and cycles require no outside intervention to keep them repeating over and over. Mechanical clocks are a relatively recent invention, though, and for the first few centuries of their existence they were anything but smooth and reliable. In fact, up through the early 1700s, writers comparing the universe to clockwork were often doing so specifically to invoke a need for active intervention, calling upon God to reach in from time to time and clean or realign some of the elements to keep everything in the world functioning properly.

The shift from regarding clockwork as balky and unreliable to considering it a paradigm for something that functions smoothly on its own was an enormous change in philosophy, driven by a dramatic leap in the technology of timekeeping that took place in the latter half of the 1600s. That leap, in turn, was enabled by developments in the philosophy of the natural world that were centuries in the making but rapidly accelerated during the same period that mechanical clocks began to emerge.

ON THE VERGE OF A REVOLUTION

Making a reliable mechanical timekeeper based on the motion of physical objects requires solving two fundamental problems: finding a motion that is sufficiently regular to keep time well, and keeping that object in motion. It's easy enough to identify physical systems that execute repeated motions that can be counted to track time, but all of them fall short in one way or another. A ball dropped from some height will bounce repeatedly, for example, but it's easy to see that each bounce takes slightly less time than the previous, and it will quickly settle down. A pendulum swinging back and forth is more reliable in terms of the time per swing, but will still slow down and stop moving over relatively short timescales.

This is analogous to the situation with water clocks that require regular refilling and sandglasses that need regular turning, but it's relatively easy to make versions of those devices that run for an hour or more. With good modern bearings and careful construction, we can make a pendulum for an intro physics lab where students can count a hundred clear oscillations before it stops. That lasts a few minutes at most, not remotely long enough to make a practically useful timepiece. Finding a way to stop clocks from "running down" requires some clever engineering, which explains why it took so long for reliable mechanical clocks to emerge.

In modern terms, we understand this running down as a dissipation of energy through the action of friction. There is some residual friction in the bearings of a mechanical clock, as surfaces slide past one another, that gradually converts the energy of the macroscopic motion of the clockwork into random microscopic motions of the atoms and molecules making it up: in other words, heat. In order to keep a mechanical clock running for a long time, it must draw on a power source to periodically refresh the oscillation, adding energy to the system. This energy must be added in a way that does not perturb the oscillation of whatever moving system is used to regulate the time, but merely sustains it against losses to dissipation.

The Chinese solution, as we saw in the tower clock of Su Song, ingeniously combined these factors, using the steady flow of water as the regulation and the weight of the water to provide the impulse needed to turn the gears of the clock. In Europe, however, the two problems were largely separated: the water clocks developed around the Mediterranean were mostly float driven and could not provide sufficient force to turn overly complicated gear trains. While marvelous water clocks were developed involving both audio alarms and visual displays, these generally used the rising water level as a trigger, not a power source. The small force provided by the rising float on the clock would trip the release of a coiled spring or a falling weight hung from a cord wound around a gear shaft to provide power to the automata. These springs and weights provided a more substantial force, but one that was rapidly exhausted.

The exact moment of the mechanical clock's invention is obscured by the fact that for a period of several centuries the same word, "horologium," was used indiscriminately for any timekeeping device. This adds an element of detective work to figuring out the history of timekeeping technology, trying to determine from context clues whether a particular horologium was a mechanical clock, a water clock, or even a sundial. For example, a record that some of the monks fighting a fire at St. Edmundsbury in 1198 CE "ran to the clock" suggests it was a water clock with a substantial reservoir. On the other hand, a note that a clock installed in 1284 CE at Dunstable Priory was placed *above* the rood screen separating the lay congregation from the altar suggests that one was mechanical rather than water driven. This ambiguity makes it difficult to figure out the chain of influences, but it seems natural for the purely mechanical clock to be a technology evolved from the weight-driven automata whose displays were triggered by water clocks. Some enterprising tinkerers whose names are lost to history realized that the substantial force generated by falling weights could be used to drive the clock itself, rather than merely a display, provided a suitable means of regulating its release could be found.

By the early 1300s, records begin to include more references to devices that are unambiguously mechanical clocks. Several clocks that claim to date from the latter half of that century survive in various churches in Europe, though nearly all of them have been modified over the centuries to make them operate more accurately than they would have when they were first built.

The earliest mechanical clocks used a "verge-and-foliot" system to both regulate and sustain the operation of the clock. The main moving part is the foliot, a long, bulky rod attached to a shaft, which twists back and forth between ticks of the clock. The audible ticking is caused by the verge, a pair of pallets on the foliot shaft, one near the top, the other near the bottom, which collide with the teeth of a crown gear that is driven by a falling weight. The drive shaft for the crown gear holds a spool of cord with a weight on the end: the pull of the weight turns the drive shaft, but this rotation is interrupted by the pallets striking the teeth of the crown gear.

We can see how this operates by considering one full cycle of the foliot oscillation. If we start with the foliot at rest and the top pallet in contact with the crown gear, the force from the hanging weight trying to turn the drive shaft delivers a force through the gear to the pallet. This force starts the foliot rotating in a clockwise direction, and as it turns, the top pallet of the verge rotates out of the way, allowing the crown gear to turn. A short time later, though, the bottom pallet slams into a tooth on the lower side of the gear, bringing both the foliot and the crown gear to a temporary stop. The driving force from the weight then pushes the foliot to rotate in a counterclockwise direction, moving the bottom pallet out of the way and allowing both gear and foliot to move until they are stopped by the top pallet again colliding with the crown gear. At which point the cycle starts over.

The characteristic "tick-tock" of a mechanical clock first appears through the action of a verge escapement. First we hear a "tick" as the top pallet collides with a tooth on the upper part of the gear, then a similar but distinct "tock" as the bottom pallet collides with a tooth on the bottom part. The verge is the first widely used mechanical clock escapement, but essentially all of them share this characteristic feature: two different surfaces striking and briefly stopping gear teeth in different positions.

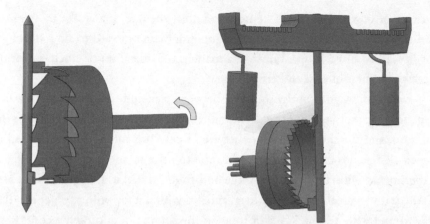

Left: A verge escapement, consisting of a shaft with two pallets that interrupt the motion of a crown gear, causing the shaft to twist back and forth, rotating by around 90 degrees. The collisions between the pallets and the crown gear teeth provide the "tick" and "tock" sound of a mechanical clock. Right: A verge and foliot clock. The vertical shaft holds the foliot (heavy horizontal bar) whose twisting motion determines the period of the clock. Hanging weights on the bar can be moved closer to the center to speed up the ticking or farther away to slow it down.

The verge-and-foliot system is an ingenious development, allowing a clock that will run as long as there is a force to drive it. The foliot is set into motion by the force on the drive shaft, and the verge escapement periodically interrupts that drive to keep the weight from falling too rapidly. The clock will keep ticking and tocking as long as the weight has farther to fall, which for a clock installed high in the tower of a medieval church can be a good long while.

Additional gearing can turn a pointer for a visual display and also trigger the ringing of bells at regular intervals, allowing tower clocks to provide time signals not just for those close enough to hear the ticking but for much of the surrounding countryside. This allows the widespread public dissemination of reasonably accurate time, and between 1300 and 1600 CE tower clocks became central features of churches and cities all across Europe. These public clocks led to profound changes in the way people ordered their lives.

As revolutionary as it was, the verge-and-foliot system leaves much to be desired from the standpoint of precision timekeeping. The time required for the foliot to rotate from tick to tock depends on not just the weight of the foliot, but its shape; a longer foliot, with its mass farther from the axle on average, will turn more slowly. This could be used to adjust the rate of a clock, by hanging additional weights from the foliot and moving them outward to slow the clock (or inward to speed it up), but it also made standardization of clock operation a tricky business of trial and error.

Even more significantly, the time between tick and tock for a foliot clock depends on the strength of the driving force: the harder the push, the faster the rotation, and the shorter the time between ticks. This means that the speed of the clock depends on the exact weight used to drive it, and even subtler effects. Experiments with the restored verge-and-foliot clock at England's Salisbury Cathedral suggest that it would keep time to within a few minutes per day if it was wound with a single layer of rope on the spool, but adding a second layer caused it to speed up. Increasing the effective radius of the spool by one layer of rope increases the torque on the drive shaft generated by the pull of the hanging weight, and thus the force delivered via the crown gear to the foliot. Carelessness or laziness in winding up the rope to drive the clock could thus make the clock run at different rates and introduce significant errors.

These issues make verge-and-foliot clocks tricky to operate and maintain. The best mechanical verge-and-foliot clocks, such as those made by the Swiss clockmaker Jost Bürgi in the early 1600s, could keep time to within about a minute per day.* Most mechanical clocks of that era were much worse though; for a long time, most mechanical clocks didn't even have minute hands, as their errors were too great. For reliable precision timekeeping to be possible with mechanical clocks, a new type of regulator needed to be developed, one that was less sensitive to changes in the driving force.

PHILOSOPHY AND THE PENDULUM

After the appearance of mechanical clocks, it took around 300 years for engineers and artisans to make a fundamental improvement over the verge-and-foliot system. What finally emerged, in the mid-1600s, was a byproduct of a revolution

---

* Bürgi was responsible for many innovations in clockmaking, most importantly inventing the remontoire, in which the main drive weight does not drive the clock directly but serves to regularly and automatically rewind a smaller weight that drives the gears. This supplies a more constant force over the full running period, making for better clock performance.

in the philosophical understanding of moving objects a thousand years in the making.

For the better part of 2,000 years, the dominant philosophical approach to physics was based on the work of the Greek philosopher Aristotle (384–322 BCE). In the Aristotelian approach to physics, the natural state of material objects was believed to be at rest at the lowest position possible. A moving object was thus displaced from its natural state, and without some continued effort to *keep* it moving, would quickly return to its proper condition. While we now recognize that this picture is fatally flawed, it took centuries of work by philosophers and mathematicians, starting in the late days of the Roman empire and continuing through the medieval period, to arrive at our modern understanding of physics, which shows that an object in motion tends to remain in motion unless it's subject to a force that acts to stop it. The signature of a force acting on an object is not motion, but a *change* in motion—speeding up, slowing down, or changing direction. An object that's moving in a straight line *has been acted on* by a force, but it doesn't need that force to continue to stay in motion. We live in a world where most of the objects we interact with are subject to friction and other forces that act to stop motion, though, so it wasn't immediately obvious that Aristotle's picture was wrong.* The most significant challenge facing Aristotelian physics, and one of the keys to moving beyond it, was how to explain projectile motion: an object thrown through the air or an arrow shot from a bow will continue in motion for a good long while without any contact with another object that could supply the force supposedly needed to keep it moving. Such projectiles will regularly remain in motion long past the time when an object rolling or sliding along the ground would've come to a stop. So, what keeps something flying through the air moving?

Aristotle's answer was to attribute the force sustaining projectile motion to the air itself. The continued motion of a projectile flying through the air was, he argued, a result of air being displaced from in front of the object and rushing around it and providing a push from behind. That push sustained the motion of the projectile long after it left the hand or bow that set it into motion. Aristotle's

---

* One of the great challenges of teaching introductory college physics, as I do on a regular basis, is breaking students out of a sort of instinctive Aristotelianism. Even future physicists and engineers tend to have an Aristotelian intuition that anything in motion is necessarily being pushed, because it seems to work well, at least on a conceptual level, for most everyday objects.

force-from-air explanation is fairly ingenious as a way to talk your philosophy out of a quandary but does not hold up well under further examination. Among other problems, it would seem to suggest that an object presenting a larger cross-section to the air ought to be propelled by a stronger force and fly farther, when in fact the opposite is true.*

While many popular myths about the history of science have Aristotelian physics reigning unchallenged all the way up to the 1600s, philosophers were questioning his ideas a thousand years earlier than that. One of the earliest good critiques of Aristotle's ideas came from John Philoponus in the mid-500s CE, who attributed the continued motion of a thrown object to some property imparted by the thrower. Philoponus called this "*vis impressa*," an "impressed force," but "impetus" is the term that eventually caught on. The idea is similar but not identical to the modern concept of momentum, which in mathematical terms is the mass of an object multiplied by its velocity.† Philoponus believed that this imparted quality dissipated naturally over time, but around 500 years later the great Muslim scientist Ibn Sina took the next step, arguing that the slowing of real projectiles was not a natural process inherent to motion but the result of air resistance. An object thrown through a perfect vacuum, Ibn Sina argued (correctly, we now know), would continue in motion forever.

As all the people involved were primarily philosophers, it was probably inevitable that discussion around the impetus theory of motion expanded to include grander and more abstract questions about the motion and stopping of objects. One of these lofty hypotheticals, concerning a tunnel through the earth, eventually spun off an example that found practical applications in the science of timekeeping.

The tunnel hypothetically stems from the Aristotelian principle that the "natural place" of solid objects is at the lowest point available: this is why the solid sphere of the earth sits at the very center of a spherical universe, surrounded by lighter elements. Heavy objects fall, in this theory, because they

---

* Galileo used a version of this argument in his *Dialogues*, asking whether an arrow shot sideways would fly farther than one shot end on, in the usual manner.

† While some of the later impetus theories had the right basic relationship between impetus and speed, they often included extra factors such as whether the motion being sustained was linear or circular, where momentum in Newtonian physics only involves motion in a straight line.

want to be at rest at the center of the universe. This raises the question, though, of what would happen if a tunnel were dug all the way through the earth, passing through the center and out the other side.* Would a rock dropped into such a hole come to a sudden stop at the precise center, that being the ideal natural location for a solid body?

While some hard-line Aristotelians argued that this would, in fact, be the case,† proponents of the impetus theory, most notably the French philosopher Jean Buridan and his follower Nicole Oresme, argued otherwise. In 1377 CE, Oresme gave a clear statement of this view of impetus, and helpfully added an everyday example:

> *For if an opening were made from here to the center of the earth and beyond and a heavy object fell through this opening or hole, upon reaching the center it would pass beyond and begin to go upward by reason of this accidental and acquired property; then it would fall back again and come and go several times just as we can observe in the case of a heavy object hanging from a beam by a long cord.*‡

The kind of strict Aristotelian approach that says an object falling through the earth should come to a stop right at the center would *also* seem to argue that a weight hung from a cord should fall only until it reaches its lowest point, then stop. This is manifestly not what happens in the real world, though, where a weight on a string swings back and forth multiple times before stopping.

A weight falling through the earth, Oresme argued, ought to do the same thing. The falling weight should cruise right on through the center point and only start to slow gradually *after* passing the center. In the absence of air resistance (if we're imagining a tunnel right through the core of the earth, we may as well imagine it containing a perfect vacuum), the weight would rise back up

---

* Contrary to myth, most medieval philosophers did not believe the world was flat but were perfectly comfortable with the idea of a spherical Earth.
† Including Adelard of Bath, who was the first to publish this example, in his *Quaestiones naturales* (written sometime around 1107–1133 CE), presented as an imaginary dialogue between a well-traveled philosopher and his curious nephew.
‡ Nicole Oresme, *Le livre du ciel et du monde* (ed. and trans. Albert D. Menut and Alexander J. Denomy). Quoted in "The scholastic pendulum," by Bert S. Hall. *Annals of Science*, 35:5 (1978): 441–462.

to ground level on the far side of the planet before stopping and falling back to return to its original point.

Oresme's hanging-weight thought experiment marks one of the first points in the history of science that a pendulum appears as a concrete example of oscillatory motion. Over the next couple of centuries, a swinging pendulum became more common as an illustration and gradually evolved into an object worthy of study in its own right.

Physics of the Pendulum

The legendary theoretical physicist Sidney Coleman famously summed up a core element of physics training in the late twentieth century by saying: "The career of a young theoretical physicist consists of treating the harmonic oscillator in ever-increasing levels of abstraction." A "harmonic oscillator" is, as the name suggests, a system that oscillates with a single characteristic frequency, and in mathematical terms, it's one of the simplest problems to set up and solve, a staple of introductory courses. The real attraction, though, is that mathematically an astonishing variety of physical systems can be made to *look* like a harmonic oscillator, provided you treat the world in somewhat abstract terms. Everything from the macroscopic vibrations of a solid object, to the microscopic motions of the atoms in a crystal, to the behavior of photons (quantum-mechanical "particles" of light) can be described by equations that have the same mathematical form. Even the most exotic and speculative theories end up being made to look like harmonic oscillators: the field of string theory stems from the realization that, when written in the proper form, the equations governing certain particle physics processes look like those governing the harmonic oscillation of a string, albeit one that's able to vibrate in more than the usual number of spatial dimensions.

The first and most concrete example in Coleman's progression through versions of the harmonic oscillator is a mass attached to the end of a spring, sliding on a horizontal surface: When the spring is relaxed, there is no force, so the mass remains at rest, but if the spring is stretched or compressed, it exerts a force that pushes or pulls the mass back toward the relaxed position of the spring. The magnitude of the force increases as the amount of stretch (or compression) increases. If you stretch the spring twice as far, the force pulling back is twice as great; similarly, you need to push with twice the force to double the compression. This is known as "Hooke's law" after Robert Hooke, who discovered the relationship in 1676 CE.*

If the spring is stretched and then released, this leads to an oscillating motion: the force pulls the mass toward the rest length causing it to accelerate, but when it gets to the point where the spring is at its relaxed length, the force is zero but the mass is moving, so it overshoots and compresses the spring, thereby increasing the force until it brings the mass to a stop. Then the process reverses, pushing the mass toward the relaxed length, overshooting, and returning to the initial stretched position. The period of the oscillation—the time to go from fully stretched to fully compressed and back to fully stretched—depends on only two factors: the stiffness of the spring (how much force you get for a given stretch), and the mass (which determines how much force you need to change its motion).

A mass on a spring is the cleanest realization of a simple harmonic oscillator, which is why it's the first system aspiring physicists encounter, but even this is an idealization of a real system whose behavior is more complex. An actual physical spring of coiled wire will depart from Hooke's law at the extremes of stretch or compression. When pushed too far, the coils come into contact with each other, and the force increases rapidly; if stretched too far, the wire will uncoil, as inevitably happens with any spring-based toy that falls into the hands of a real child. As long as you keep the displacement small, though, Hooke's law is an excellent approximation and accurately predicts the oscillating motion of a mass on a spring.

The next system in Coleman's progression is a simple pendulum, consisting of a compact mass swinging back and forth at the end of a very light string. A pendulum deviates from the ideal harmonic oscillator more rapidly than a spring does, but like the mass on a spring, it is an excellent approximation with appealingly simple properties, provided the swing is not too large. From the point of view of a physicist in the twentieth century, it's surprising, bordering on shocking, that the pendulum as a physical system was not seriously examined until nearly 1600 CE.

---

* Hooke published his result as a crude anagram: "ceiiinosssttuu" which unscrambles to "*ut tensio, sic vis,*" Latin for "as the stretch, so the force." This allowed him time to build on his discovery without competition but established his claim to priority; these kinds of shenanigans were regrettably common in the 1600s.

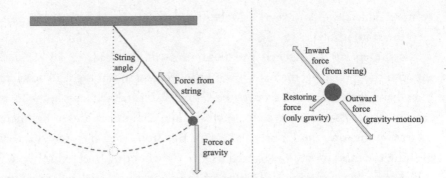

Simple pendulum on the left, showing the forces that act and the angle of the string. On the right, the situation from the point of view of the bob, with inward force from the string and outward force caused by gravity and the motion of the pendulum, which exactly cancel each other out. This leaves a restoring force, which depends only on gravity and the string angle, pulling back toward the center.

The simplicity of the pendulum comes from the fact that the pendulum bob at the end of the string is acted on by only two forces: the constant downward pull of gravity and an inward pull from the string. While the force of gravity is constant in both magnitude (determined by the mass of the bob) and direction (gravity always pulls straight down, by definition), the force from the string varies, taking on whatever magnitude it needs to keep the bob on track. This variation should be familiar to anyone who's ever ridden a playground swing. At the turning points of the swing, the string force is minimal, and when swinging in a very large arc, the chains will go slack for an instant, while at the bottom of the swing the force reaches a maximum that is greater than the weight of the rider, causing a "heavy" sensation at that point in the arc.

The varying strength of the string force and the changing angle between it and gravity make this seem like a very difficult problem when considered from the perspective of an observer standing next to the pendulum and considering the forces in terms of fixed vertical and horizontal directions. To such an observer, the bob itself is always moving along both up-down and left-right directions, and both the up-down and left-right components of the force are constantly changing.

One of the features that makes pendulum motion an attractive problem for introductory physics courses is that the situation can be radically simplified by a change of perspective. If we think about the motion not from the perspective of an outside observer but from an observer swinging on the bob, the problem becomes much more tractable. From the viewpoint of the rider, we can describe the motion not in terms of vertical and horizontal distances, but in terms of the angle the string makes with the vertical and a distance in and out along the string. Since in ordinary pendulum motion the bob does not move either in or out along the string, that component never changes, and we only need to worry about a single value, the angle, to describe the position. When we look at the forces in this picture, we find a similar simplification: the inward force from the string must be exactly canceled by the outward force associated with the moving bob.* The only force we need to worry about for finding the period of the pendulum's oscillation is the force component directed back and forth along the arc.†

The mathematical form of the back-and-forth force can be determined from a bit of trigonometry, and depends on the weight of the bob (its mass multiplied by the strength of gravity) and the sine of the angle. This provides a restoring force, one that always points back toward the center, increasing as the angle increases, and leads to oscillating motion.

In an ideal harmonic oscillator, the force would follow Hooke's law and always increase linearly—doubling the angle would always double the force. That's not strictly true for a pendulum, where the force has a maximum value: it can't be bigger than the weight of the bob. When the string is pulled to an angle of 90 degrees, the bob starts out simply falling straight down; well before that, though, the rate of increase in the restoring force has fallen below what we expect from Hooke's law. As long as the oscillation is relatively small, though, the pendulum behaves very much like a harmonic oscillator, oscillating back and forth with a single well-defined frequency.

---

* This is a combination of the gravitational force on the bob and a fictitious "centrifugal force" associated with the motion of the bob on the circular arc, which depends on the speed and the length of the string; this is what causes the "heavy" feeling at the bottom of a playground swing.

† The magnitude of the inward force that must be supplied is, of course, of critical importance to, for example, designers of playground equipment, but it has no impact on the period of the oscillation, which is what matters for timekeeping.

In the case of a pendulum, the force that's pulling the bob back toward the center originates from gravity, so it increases as the mass of the bob increases. But we also know from Newton's laws of motion that the force needed to move an object increases as the mass increases, and these two effects cancel each other out. The motion of the pendulum, and thus the frequency of its oscillations, does not depend at all on the mass of the bob: the oscillation frequency depends only on the length of the string and the strength of gravity, which is pretty close to constant near the surface of the earth.

These two features of pendulum motion make it exceptionally attractive as a regulator for a mechanical clock. Unlike the verge and foliot, whose period of oscillation depends on both the mass of the foliot and the strength of the pushing force, a pendulum is relatively insensitive to both of those factors, making a pendulum clock easier to make and calibrate, and more reliable over long periods of operation. It took a long time for anyone to think of it—280 years passed between Oresme's passage quoted above and the first pendulum clock in 1657 CE—but once established, it spread like wildfire.

From Philosophy to Product: Engineering Pendulum Clocks

The first working pendulum clock was designed by Christiaan Huygens in the Netherlands and built by Salomon Coster in 1657. Huygens was a key figure in the burgeoning scientific revolution, making groundbreaking contributions to optics, astronomy, the mathematics of probability, and other fields. His most notable contribution for the purpose of this book was to place the emerging science of physics on a firmer mathematical footing, establishing clear formulae governing the behavior of many kinds of moving objects, and reanalyzing problems in a more rigorous way.

Huygens eventually produced a definitive study of the pendulum, his 1673 book *Horologium Oscillatorium*; among the many problems addressed was isochrony. Huygens showed mathematically that the circular arc of a simple pendulum, as described above, does not, in fact, produce isochronous motion. He went even a step further, identifying the proper shape of a curve that does: particles sliding down the surface of a cycloid (the shape traced by a point on the rim of a rolling circle) will take the same time to reach the bottom regardless of where they begin. Compared to the circular arc of a simple pendulum, a cycloid is somewhat steeper, and Huygens worked on producing a truly isochronous pendulum by

THE (EMBELLISHED) EXPERIMENTAL PENDULUM

One of the first systematic investigations of the motion of a pendulum as a physical system was carried out by Galileo Galilei around 1600, but shared primarily with his patron of that time, Guidobaldo del Monte. In the course of this work, Galileo conceived the idea that the pendulum is "isochronous," taking the same time to complete one back-and-forth swing regardless of how far the bob is drawn back, an idea he would return to many times. He tested this using a water clock and a balance scale, weighing the water collected during swings of different lengths and comparing them. He also used free fall as a timer: having previously established his belief that all objects fall at the same rate, he would drop a weight onto the floor at the same time he released a pendulum to swing into a vertical board, and measure the height needed for the sounds of the two impacts to be simultaneous.

Galileo also observed the dependence of a pendulum's period on its length, though this was initially useful to him largely for rhetorical reasons. The context for his first published reports about pendulum motion was the ongoing argument about heliocentric vs. geocentric models of the solar system. Galileo was promoting a Copernican model of the solar system, with the planets including the earth orbiting the sun, and in his (in)famous 1632 *Dialogue Concerning the Two Chief World Systems** used the pendulum as an analogy, arguing that it was natural for the inner planets to move more rapidly through their orbits than the outer, in the same way that a longer pendulum takes a longer time to complete its swings.

The rhetorical character of Galileo's work is also shown by his tendency to . . . embellish his results. In the *Dialogue* and the 1638 *Discourses and Mathematical Demonstrations Relating to Two New Sciences*, he made grand claims about the isochrony of the pendulum, for example, writing in the *Discourses* that if two people start counting the swings of pendulums set swinging with very different amplitudes, "they will discover that after counting tens and even

---

* This is the book that got him in trouble with the Catholic Church, though contrary to myth this was more about politics than science: the church was less bothered by the use of a Copernican model for the solar system than by the fact that both in the text and through the act of publishing it when he had promised not to, Galileo's book made the pope look ridiculous.

hundreds they will not differ by a single vibration, not even by a fraction of one." In fact, it is very difficult to get a pendulum of the sort used by Galileo to swing through even 100 oscillations, and if you do, the very real difference in the period between a large amplitude and a small one (which happens because the pendulum is not an ideal harmonic oscillator) will become apparent. His contemporaries, including René Descartes, Marin Mersenne, and Giambattista Riccioli, noted the dramatic discrepancy between Galileo's claims and their observations, with Mersenne going so far as to say he doubted Galileo had done the experiment.

In fact, Galileo's experiments timing pendulum swings with his water clock had shown a difference in the time for two different arcs, one released from 10 degrees, the other swinging through a full 90 degrees. As a percentage of the time for the swing, the difference came out to about 9.4 percent, whereas a modern analysis (taking into account the deviation from the ideal harmonic oscillator) would predict a difference of 10 percent. Galileo's error was not in the experiments but in his interpretation: his measurements clearly showed that there were differences in the time required for pendulum swings of differ-ent amplitude, but he interpreted these as the result of "accidents": interac-tions such as friction and air resistance that prevented his real pendulum from behaving as it would in an ideal case. He firmly believed that the ideal pendulum was isochronous, and only *appeared* not to be because of annoyingly inescapable imperfections in his real-world apparatus.

When writing about these results for popular consumption, he took the opportunity to "clean things up" a bit. In modern terms, Galileo was a cross between a scientist and a salesperson: he carried out and reported on experi-ments, but he did so in service of selling a particular view of the world,* and he did not shy away from slanting his presentation of experimental results in ways that made the case for his ideas seem more persuasive. Ironically, he was in a sense engaged in the same sort of behavior that modern myth attributes to his Aristotelian rivals: pure reason told him that the universe ought to behave in a

---

* And also selling himself: in Galileo's day, most scientists were supported by wealthy patrons, and he worked very hard to ingratiate himself with the most wealthy and powerful men in Italy, so as to secure more prestigious positions for himself and his family.

certain way, and when experimental reality occasionally failed to live up to that, he presented an idealized version that did.

In many ways, this idealizing impulse pushed him in the right direction, as in his discussion of the behavior of falling objects. In the presence of air resistance, denser objects do, in fact, fall faster, contrary to Galileo's assertion that all objects fall at the same rate.* But this *really is* a matter of "accidents" upsetting the result: if we take away the effects of air, the rate of free fall genuinely is independent of mass.[†] In the specific case of the pendulum, his instincts led him astray, because the mathematical approach that could show that the period really does depend on amplitude had not yet been invented.

While he was wrong about the pendulum being perfectly isochronous, a frequency shift of only a few percent through wide ranges of amplitude was very good by the timekeeping standards of the mid-1600s. In his final years, Galileo became the first to propose using a pendulum to regulate the ticking of a mechanical clock. His former student and later biographer Vincenzo Viviani wrote that in 1641, Galileo had the idea to use a pendulum in place of a foliot, with the regular pushes from a weight-driven gear serving to keep the pendulum swinging back and forth. By this time, though, Galileo was more or less blind, so while Viviani did make a drawing of the scheme, and he and Galileo's son, Vincenzo, worked on a model, they never managed to build a working clock before Galileo's death in 1642.[‡]

---

* Riccioli and others also demonstrated this.
[†] This happens for the same reason that the pendulum period is independent of the mass: the gravitational force on a heavy object is greater, but a heavy object requires a greater force to accelerate it, and the two effects cancel out. This was spectacularly demonstrated during the Apollo 15 mission, when Commander David Scott dropped a hammer and a feather on the airless surface of the moon, showing that they fell at the same rate.
[‡] Vincenzo Galilei did not long outlive his father, dying in 1649, and whatever clock models he made were destroyed around the time of his death.

adding curved "cheeks" to either side to effectively shorten the string for larger swings, forcing the bob onto a cycloidal track. The additional friction added by the string impacting the cheeks made this impractical as an actual timekeeping device, though, so it remained something of a mechanical curiosity.

While the simple pendulum is not perfectly isochronous, its motion is less sensitive to small changes in the driving force than that of a foliot, so it does make a good basis for a mechanical clock. Huygens's initial pendulum clocks used a verge escapement, with the pendulum swinging on an axle with two pallets that impacted the teeth of a crown gear driven by a falling weight or a coiled spring. At each extreme of the pendulum's swing, the pallet on that side would be pushed into the path of the gear tooth; their impact would temporarily halt the turning of the gear and also provide a small kick to the pendulum, causing it to swing back in the other direction. These kicks keep the pendulum swinging despite the resistance caused by friction on the axle. The force involved can also be relatively small—unlike in a verge-and-foliot clock, the driving gear only needs to provide a small impulse to make up for the energy lost to friction. The drive doesn't need to supply enough force to stop and reverse the pendulum swing; gravity does that, automatically. Since the overall driving force is smaller, the range of possible variation in the force also shrinks, and with it the possible timing errors.

Huygens's pendulum clocks represented a great leap forward in mechanical timekeeping: his initial clock could keep time to an accuracy of about a minute per day, matching the very best verge-and-foliot clocks, and later improvements reduced the errors to around 10 seconds per day. This design still left room for improvement, though: in particular, the verge escapement was a source of trouble, as it required the pendulum to swing through a large arc to get the pallets clear of the gear teeth. Typical verge escapements use pallets that require a swing of 90 degrees or more (i.e., 45 degrees in each direction), and such large swings are where the nonideal aspects of a real pendulum are most noticeable. Pendulum clocks are inherently less sensitive to changes in driving force, but those with a verge escapement are about as bad as they can be in terms of the "circular error" caused by variations in driving force.* The wider the arc, the less ideal the

---

* This problem is magnified in smaller clocks (like the one in our dining room), which are driven not by the constant pull of a falling weight but a tightly wound spring whose force weakens as it relaxes.

pendulum behavior, so a change from a swing of 45 to 46 degrees changes the length of a tick by much more than a change from 5 to 6 degrees.

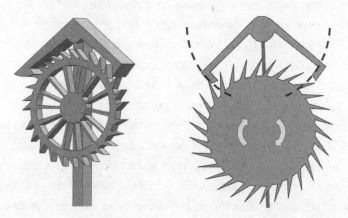

Anchor and deadbeat escapements. In a basic anchor escapement (left) the pallets that interrupt the turning of the drive gear are at the ends of long arms, so the pendulum needs to swing through only a small angle to move them in and out of the way. In a deadbeat escapement (right) the pallet faces are curved in a way that locks the gear in place as each tooth slides across the pallet, eliminating the problem of recoil in the gears.

The simplest way to avoid circular error, then, is to limit the swing of the pendulum to a smaller range. This was made possible with the invention of the "anchor escapement" very soon after the first pendulum clocks were made.* The anchor escapement takes its name from two arms extending off the pendulum axle, shaped something like the arms on a ship anchor. The pallets that interrupt the motion of the driven gear are located at the ends of these arms, allowing them to dip in and out after swings of only a few degrees in either direction. This keeps the swing in the small-angle range, where the behavior is closest to ideal, and it allows the use of a long and heavy pendulum, which further reduces the sensitivity to small perturbations of the driving force.

One final issue that was mitigated by escapement design is the problem of recoil in the gears, where the impact between pallet and gear could cause

---

* The anchor escapement was probably invented around 1657 by Robert Hooke, though credit for it was sometimes claimed by clockmaker William Clement, who began selling clocks with an anchor escapement around 1680. The two squabbled over priority for some time, which was pretty typical of Hooke's career.

the driven gear to slip backward slightly. This would lead to increased stress throughout the gear train, with the resulting wear and tear increasing friction within the works and requiring more frequent maintenance. This recoil was eliminated by the use of a "deadbeat escapement," developed by the astronomer Richard Towneley and the clockmaker Thomas Tompion in the 1670s. The deadbeat is a variant of the anchor escapement with curved faces at the ends of the arms that match a circle centered on the axle for the pendulum. One or the other of these is in contact with the teeth of the drive gear for most of the swing, locking it in place as the tooth slides across the "dead face" of the escapement: the tooth slides along the curve but doesn't require the gear to turn in the process. The gear turns only in a brief period when the pendulum is near the center of its swing, when the dead face moves out of the way and the gear tooth pushes against the "impulse face" at the end of the arm, which is angled to convert the push from the tooth into a push on the pendulum.

Deadbeat escapements became widely available around 1715 in clocks made by English clockmaker George Graham, and they quickly became the standard for accurate pendulum clocks. An antique "longcase" or "grandfather" clock almost certainly tracks time using a pendulum with a deadbeat escapement, powered by a slowly dropping weight (cranked back up to its starting height daily or weekly, depending on the clock). These became nearly ubiquitous in the 1800s, making them a signature background element in any story set in an old house. All across Europe, large public clocks running on verge-and-foliot systems were converted to pendulum clocks, often adding minute hands in the process, and civil timekeeping took a huge leap forward.

Graham's best clocks were accurate to about a second per day, a level of accuracy at which other subtle issues start to come into play, such as the effect of temperature. Most materials expand when heated, so an increase in the temperature of a pendulum clock will make the pendulum just slightly longer. This longer pendulum will tick slightly more slowly, throwing the clock off. The change in length will be tiny, but it doesn't have to be large if your clock is otherwise good enough: for a meter-long pendulum, an increase in length of about 1/40th of a millimeter (one-quarter of the diameter of a human hair) is all it takes to throw the clock off by one second per day. To get that kind of length change from a pendulum made of brass (a common material used to make clocks in the 1700s) would require a temperature change of less than two degrees Celsius.

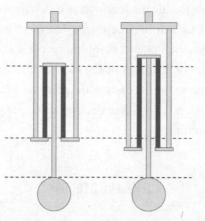

The gridiron pendulum invented by John Harrison, with a three-segment shaft made of two different metals, one that expands more than the other with an increase in temperature. The lengths of the shafts are chosen so the greater expansion of the short middle segments exactly compensates for the smaller expansion of the longer end segments, as seen on the right.

Such high temperature sensitivity is obviously a huge problem for attaining second-a-day accuracy, but Graham found a clever way to compensate. The length that matters for determining the period of a pendulum is the distance from the pivot point to the center of mass of the pendulum, which depends on the distribution of mass. An increase in the overall length can be compensated by somehow shifting some of the mass closer to the pivot, so that's what Graham did: he added a sealed tube of mercury to the bob of his pendulum clocks.* As the temperature increased, the mercury in the tube would expand and push the liquid higher up the tube, just as in a thermometer. Moving some heavy liquid mercury up the shaft compensated for the lengthening of the shaft pushing the bob farther away, hence allowing Graham's clocks to tick reliably across a wider range of temperatures.

Another ingenious solution to the problem of thermal expansion, one better suited to manufacturing in large numbers, was the "gridiron pendulum" developed by the clockmaker John Harrison (whom we'll talk about more in Chapter 10), who constructed a pendulum consisting of a zigzag pattern of

---

* As noted in the previous chapter, mercury is a very heavy metal that is liquid at room temperature.

alternating rods of two metals that expand at different rates as the temperature changes. These were arranged in such a way that the relatively small expansion of the two sets of iron rods would combine to push the bob down by some amount, while the larger expansion of the one set of brass rods pushed it back up by exactly the same amount. Harrison's gridiron pendulum kept the total length constant, to such a degree that the best of his longcase clocks was accurate to one second a month.

TIME AND GEODESY

With circular error limited by using escapements that required only a small swing, and thermal errors limited by the choice of materials, the best seventeenth-century pendulum clocks worked at an unprecedented level of precision. This logically turned them into instruments for measuring the earth itself, thanks to the one other factor that influences the period of a simple pendulum: the strength of gravity. While gravity pulls all objects at any point on the earth downward at the same rate regardless of their mass or composition, the resulting acceleration of around 9.8 meters per second per second is only approximately constant across the surface of the earth.* The variations are small, but readily detectable with a good pendulum clock.

This was dramatically demonstrated on a 1672 expedition that sent the French astronomer Jean Richer to Cayenne in French Guiana, just a few degrees above the equator. The primary goal of this expedition, planned in consultation with Jean-Dominique Cassini and Jean-Félix Picard back in France, was to help determine the scale of the solar system by making a series of observations of Mars, which would be at its closest to Earth in the fall of that year. Timing is an essential element of these measurements, so Richer brought high-quality clocks with him, which he calibrated against local astronomical observations. In the process of making these measurements, he found that his clocks from Paris ran slow by more than two minutes per day.

---

* Acceleration is a change in velocity, so it's expressed in units of speed (meters per second) per time. An object dropped from rest will increase to a speed of 9.8 m/s (a reasonably fast running pace) in the first second of its fall.

Another of Richer's tasks, closely related to the calibration of his clocks for the astronomical measurements, was to determine the length of a "seconds pendulum," which takes exactly two seconds to complete an oscillation (when used in a clock, this allows the driving gear to move forward once per second). The length of a seconds pendulum in Paris was known to be 99.353 cm in modern units, and this was being proposed as a standard for defining the length of a meter. Scientists in Paris hoped that the seconds pendulum would tie the standard of length to a universal constant that could be measured by anyone on Earth. Unfortunately, Richer found the length of a seconds pendulum in Cayenne to be a bit less than 3 mm shorter than the length in Paris.

The only two factors that determine the period of a pendulum are its length and the strength of gravity. The slow running of Richer's pendulum clocks and his measurement of a shorter seconds pendulum both point to the same issue, namely, that gravity is very slightly weaker in Cayenne than in Paris. Many scientists questioned Richer's results when they were reported, particularly Huygens, who was nursing a grudge against Richer over a test voyage that showed pendulum clocks didn't work well at sea.* Follow-up expeditions found similar results, though: a seconds pendulum in the tropical latitudes was slightly shorter than one in Europe.

Among those who believed Richer from the start was Isaac Newton in England, who was then in the process of developing his theory of gravity. Newton recognized that this discrepancy in the length of a seconds pendulum could be explained by the earth bulging outward slightly at the equator. This equatorial bulge would put the surface at the equator slightly farther away from the center of the earth, thus slightly reducing the strength of gravity at low latitudes. Richer's voyage is now remembered as a milestone in the field of geodesy, one of the first uses of clocks as a tool to measure the shape of the earth.†

---

* We'll talk more about this in Chapter 10. Huygens felt that the clocks had only failed because Richer had not properly cared for them, but several decades of trials by others bore out the conclusion that a simple pendulum clock does not keep accurate time on the deck of a wave-rocked ship.

† This was not without controversy in the early days, as Richer's boss, Cassini, was a proponent of a model where the earth is instead stretched at the poles. The bulging at the equator can also be explained by Newtonian physics, though, as a result of the rotation of the earth, so it was accepted relatively quickly.

Richer's measurements and the development of temperature-compensated pendulum clocks by Graham and Harrison are another turning point in the history of timekeeping. They represent the juncture where human-made timekeepers reached a level of precision sufficient to call into question phenomena previously believed to be absolute and universal. Prior to Richer's measurements, there was a belief that the constancy of gravity would produce an absolute and universal standard for time: anybody with sufficiently good tools to make a pendulum of the proper length could make a clock that would tick exactly once a second. Richer's measurement in Cayenne showed that this belief was premature, as applied to pendulum clocks, but both the pursuit of a universal standard for time and the battle to identify and correct for tiny systematic effects have remained central themes of timekeeping and clockmaking for the last 350 years.

Richer's long voyage (and Huygens's beef with him) point to another problem that emerged during these years of emerging global empires: the demand for timekeeping was not and could not be confined to solid ground. The longcase pendulum clocks built by Tompion and Graham and Harrison were marvelously good timekeepers *on land* but failed dismally when operated on the unsteady deck of a sailing ship. As European nations began to expand around the globe, the problem of determining the time while at sea became a critical one. Finding a solution required tremendous mechanical ingenuity as well as careful observations and calculations in astronomy, which was going through a revolution of its own in the 1600s. Before we move on to the next great developments in mechanical timekeeping, we'll first turn our gaze back to the skies.

Chapter 7

HEAVENLY WANDERERS

I n a much-loved 1996 episode of *The X-Files*, a man who may have witnessed a UFO landing is startled when a large black Cadillac suddenly pulls into his garage. The window of the car rolls down to reveal former pro wrestler and sometime politician Jesse Ventura, who launches into a threatening monologue, beginning with: "No other object has been misidentified as a flying saucer more often than the planet Venus."

Venus is the third-brightest object in the sky, after the sun and moon, and the brightest example of a category of objects visible in the night sky that we haven't yet discussed, the planets. There are five planets that can be seen with the naked eye—from brightest to faintest, Venus, Jupiter, Mars, Mercury, Saturn—and they sort of mix characteristics of the moon and the stars. The planets appear much like stars, as tiny bright points of light, but like the moon they shift positions over time. This is reflected in the name "planets," derived from the Greek name for this collection of objects, "*planetes asteres*" or "wandering stars."

While Ventura's character in his *X-Files* cameo is suggesting it for nefarious purposes, Venus is in fact an excellent candidate for misidentification as a UFO, partly because it is so bright, but mostly because it sometimes shows up in unexpected places. Venus, like all the planets, moves about in complicated ways, appearing in different parts of the sky at different times, and even disappearing for long stretches of time. We know more or less where and when to expect the moon, but Venus can catch people off guard, appearing as a brilliant spot of light in a place where none is supposed to be.

The complex motion of the planets makes it difficult to use them as a means of timekeeping—they don't reliably appear in a particular place in a particular season—and as a result many cultures have not attached much significance to them. For some cultures, though, the motion of the planets posed a challenge to the predictive function of timekeeping, a deviation from the comfortingly regular "ticking" of the motion of the sun and moon. This made them a nearly irresistible problem to be solved by careful observations and recordkeeping over a span of centuries. All this astronomical effort would in time lead to a revolutionary reassessment of the laws that govern the behavior of the universe, and of our place in it.

THE MORNING AND EVENING STAR

While its brightness ensures that every human culture has noted the motion of Venus, the ancient Maya seem to have attached greater significance to it than most. Venus was associated with a warlike deity, and Venus iconography is found both on major temples and in a famous mural at Bonampak commemorating a military victory. The Venus symbols, together with the recorded date near the heliacal rising of the planet, seem to suggest that the victors attributed their success to observations of Venus that let them know the most auspicious time to attack.

Tracking and predicting the motion of Venus takes up a significant amount of space in the Dresden Codex (described in Chapter 4). Six of the 78 pages of the codex are given over to tables relating to Venus, including lavish illustrations of the Venus god in five different costumes. And these tables are remarkably successful at predicting the appearances and disappearances of Venus over a period of several centuries, a great testament to the care and precision of Mayan astronomers.

As the very brightest of the planets, and one of the fastest moving, Venus makes an excellent place to start considering planetary motion. While its motion seems complex to a casual observer, once you get past the fact that it moves at all and settle in to observe its motion over a long period of time, a clear and relatively straightforward pattern emerges. First and foremost, Venus turns out to be visible only in the few hours immediately before sunrise or after sunset, from which it picked up the incorrect but useful monikers "morning star" and "evening star."

There is also a distinct pattern to Venus's position relative to the sun, if you track it just at the moment of sunrise or sunset: it first becomes visible just barely above the horizon before dawn, then moves higher in the sky on successive sunrises over a period of months. After reaching a peak where it rises a few hours before the sun, it begins to move back down the sky, returning to "just barely above the horizon" after about 263 days. Then there are about 50 days when Venus isn't visible at all, after which it reappears on the horizon immediately after sunset. For the next 263 nights, Venus is visible in early evening, following a similar pattern of moving up the sky and then back down. Then it's invisible for eight days before returning as the morning star. Then this roughly 584-day cycle repeats.

If you track the position of Venus exactly at sunrise or sunset over a period of many years, you will find a complex pattern that involves shifting north and south as well as up and down the sky. Given enough time, though, a careful observer will find a regularity to these patterns as well: for both morning star and evening star periods, there are five distinct shapes that Venus traces in the sky, and these repeat in a particular order. Five repetitions of the 584-day Venus cycle is almost exactly eight tropical years, so every eighth year of your long-term program of Venus-watching will find the planet rising at the same spot on the horizon, in the same season of the year.

The Mayan astronomer-priests who made the Dresden Codex recognized these patterns and turned them into a set of predictive tables. The five main pages of the Venus table contain lists of intervals and date names in the *tzolkin* calendar, reporting the motion of Venus. A sample passage (translated by Aveni) reads:*

> And then on the day named 10 Cib, . . . Venus . . . disappears in the east . . . having been seen for 236 days . . .

---

* Anthony F. Aveni, *Stairways to the Stars: Skywatching in Three Great Ancient Cultures*. John Wiley & Sons, Inc., 1997.

> *And then on 9 Cimi, in the west, Venus reappears, from the north, having been*
> *absent 90 days*
> *And then on 12 Cib . . . moving to the south, Venus disappears in the west, having*
> *been seen 250 days*
> *And then on 7 Kan, in the east, Venus reappears from the south, having been absent*
> *8 days*

This pattern of intervals—236, 90, 250, 8 days—repeats over and over, with each of the eight-day disappearances followed by one of the pictures of the Venus god. After moving across the whole five-page span, the table loops back to the start, and the five-part cycle repeats. There are 13 rows in the table, totaling 37,960 days, so the full table covers more than a century of Venus cycles.

The intervals assigned by the Maya to the stages of Venus's cycle also total 584 days, the whole number closest to the period of the five-part cycle of Venus. This is slightly off the true period, though, so at the end of a pass through the full table, the predicted date for the reappearance of Venus in the morning sky would be off by about 5.2 days from the actual rising. The Maya were aware of this, so the tables come with a "user's guide" on the preceding page, which instructs the reader to deduct four days before starting over; this brings the prediction more in line with reality and also allows the reentry of the table to occur on the *tzolkin* day 1 *Ahau*, a combination associated with Venus. On a longer timescale, there are instructions for an occasional "double correction," subtracting eight days, which again would allow reentry on the auspicious date of 1 *Ahau*.*

The Dresden Codex tables include a notional start date in the Long Count format of 9.9.9.16.0, corresponding to February 6, 623 CE. That date is actually badly off from the true heliacal rising of Venus, suggesting it reflects a bit of backdating on the part of the priests drawing up the table. Working through

---

* The way these corrections synch up with the *tzolkin* cycle points again to the social dimension of time and how the Maya were balancing different factors in their calendar system. It may also play into the puzzling difference in the shorter intervals within the cycle—the Dresden codex value of 236 days for the length of the first morning star phase is distinctly shorter than the modern interval of 263 days. This may reflect an attempt to synchronize the Venus tables with some other cycle that the Maya regarded as critically important; we simply don't know.

the full table twice (including the corrections) lands on the date 10.5.6.4.0 (November 20, 934 CE), which perfectly matches the date of Venus's heliacal rise, suggesting that this may be the true date when the table was implemented.* Following through the table from there and applying the recommended corrections gets to 11.5.2.0.0 in the Long Count (December 22, 1324 CE) with an error in the predicted date of Venus's reappearance of only 3.2 days. By this time, the classical Maya civilization had long since collapsed, and there were no more astronomer-priests to make further tweaks to the table.

The illustrations of a spear-flinging Venus god remind us that the Mayan codex was fundamentally a divinatory device, not a scientific treatise: the pictures and surrounding text describe omens associated with the five different stages of the Venus cycle.† While their purpose was very different from that of modern Western astronomers, the overall accuracy of the heliacal rise predictions stands as a testament to the care and patience of Maya astronomers.

The Venus tables of the Dresden Codex are perhaps the most impressive accomplishment of Mayan astronomy, but they were not a singular obsession. A similarly organized set of tables elsewhere in the document predict phases of the moon, with notations indicating the times when eclipses are most likely. Another six-page table tracks the motion of the planet Mars, using repeating cycles similar to those of the Venus tables. The other reasonably intact surviving codices, in Paris and Madrid, also include astronomical tables, though not so elaborate as those in the Dresden Codex. The ubiquity of such tables, and of astronomical symbols and alignments in temple architecture, suggest that careful watching of the skies was a central activity of the Mayan religion, and the success of such tables at predicting the motion over a period of centuries shows that they were first-rate naked-eye astronomers.

---

* This start date is particularly championed by the archaeologist Floyd Lounsbury, who was the first to check the Dresden table against modern calculations of the Venus cycle.

† Mostly bad omens: "Woe to the moon," "Woe to man," and "Woe to the maize god" are among the signs that have been translated. The Maya do not seem to have been a cheerful people.

YOU SPIN ME RIGHT ROUND

While the Dresden Codex tables testify to the empirical success of Mayan astronomy, we don't know what physical model, if any, they used to understand the motion. With our modern knowledge of the solar system, though, we can explain all the features of the Venus cycle in a straightforward way by considering the orbital motion of the two planets.

Venus orbits closer to the sun than Earth (the radius of its orbit is about 72 percent that of Earth's), which is why Venus is never seen high in the sky at midnight. The time required for Venus to complete an orbit is also shorter than an Earth year, just 225 days. We can combine these facts to make a simple model of the motion of Venus as seen from Earth by looking at their relative positions.

Orbits and Angles

Of course, we can't move easily among the stars and planets to measure absolute distances between objects in the sky, so our measurements need to work just on what we can see. This means that position measurements in astronomy always come down to measuring angles: how far and in what direction do we need to turn our eyes (or telescope) to move from looking straight at the object of interest to be looking straight at some reference object. For really fine measurements, the reference objects are distant stars, but for the purposes of understanding the cycles in the Dresden Codex, we can just think about the angle between Venus and the sun just at the instant of its rising or setting. We can determine this by

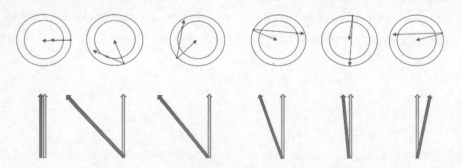

Seen from the earth, Venus is initially in line with the sun, but as Venus pulls ahead the two separate, with Venus rising as the "morning star." They then come back together as Venus moves behind the sun, and then move apart in the opposite direction, with Venus as the "evening star."

tracking the positions of the two planets in their orbits and drawing arrows to show the line of sight from the earth to Venus and the earth to the sun.

If we begin with Venus directly between the earth and the sun, Venus quickly moves ahead of the earth, and the angle between Venus as seen from the earth and the sun (from the perspective of an observer on the earth) increases rapidly to the left in the diagram. This corresponds to the heliacal rise of Venus as the morning star, followed by a rapid climb: each morning at the instant the sun peeks above the horizon, Venus is a little higher in the sky than it was the day before. As Venus continues to move ahead, though, its orbit carries it toward the opposite side of the sun from the earth; during this phase, the angle between Venus and the sun decreases, and Venus moves back down the sky. As Venus passes behind the sun, it's invisible from the earth for some time before reemerging on the other side as the evening star. The cycle of increasing and then decreasing angle repeats, until the two are once again directly in line; then the cycle repeats.

The two periods of visibility, in the morning and evening, are roughly equal in length, but the two periods when Venus is lost in the glare of the sun have very different lengths. When the two planets are on opposite sides of the sun ("superior conjunction" in astronomy jargon), both planets are moving in a direction that tends to keep the sun between them, like two people in a slapstick comedy chasing each other around a table. Venus's faster motion eventually makes it pull ahead, but the period in which it's hidden lasts several weeks. When they're on the same side of the sun ("inferior conjunction"), Venus's faster orbit takes it away from the sun in a matter of a few days. This is why the time between Venus's disappearance from the morning sky and its reappearance in the evening lasts 50 days (90 in the Dresden Codex tables), but its disappearance from the evening sky and return in the morning lasts only eight days.

Reversals and Angles

The Dresden Codex tables are remarkably successful for what they do, but have the relatively modest goal of predicting only the rising and setting dates of Venus. A much tougher problem is to predict the finer details of its motion *between* those dates, as it moves up and down the sky, and north and south. And Venus is constrained in its motion in ways that the other visible planets are not, making the others even more difficult to predict.

The fundamental problem that makes planetary positions so difficult to predict is that while the visible planets also move through the constellations of the zodiac, unlike the sun and the moon, they sometimes move backward. The sun progresses through the constellations in a particular order, sometimes moving a little faster, sometimes a little slower, but always in the same direction. The moon does the same thing, only over the course of a month rather than a year. A visible planet like Mars, though, will *mostly* shift slightly to the east from one night to the next, but every two years or so Mars temporarily reverses its nightly eastward drift and moves *west* from one night to the next. After a period of a few months, the westward motion stops, and the planet resumes moving east, night after night.

This "retrograde motion" happens with all the visible planets at different intervals (once every 26 months for Mars, but about once a year for Saturn) and for different lengths of time (around 72 days for Mars, 138 for Saturn). Explaining why this happens poses a major problem for attempts to predict the motion of the planets in detail, and it's challenging enough that some cultures seem not to have bothered. The Maya, however, were not such a culture. The Mars tables in the Dresden Codex are organized in a way that clearly highlights the retrograde motion periods, suggesting that they were well aware of the phenomenon. The five different Venus gods in the codex, each with their own omens, also suggest that the Maya recognized the different patterns in Venus's motion.*

While we don't have detailed records of Mayan observations of Mars and Venus, we can at least say a few things about the general process of making such observations, which would be the same everywhere. Again, measuring positions in astronomy is a matter of measuring angles, and when we want to measure the exact position of a planet in the sky, we require at least two angle measurements, but three or more are even better. If you know the angular separation between the object you're interested in and three nearby bright stars, you can do a bit of geometry to determine the exact position on the sky of the object you're

---

* Venus does have periods of retrograde motion, but these occur when the planet is in conjunction with the sun, and as a result they are difficult to see.

interested in.* Each angle defines a circle of that radius about its central star, and those three circles will intersect at one and only one point on the sky.†

Three angles and three reference stars will uniquely define the position of a planet *relative to those three stars*, but if we want to predict the position of a planet that moves through the full zodiac over time, we need a way to connect measurements using different reference stars. This requires a good stellar atlas, giving the positions of the brightest stars relative to each other, across the whole of the sky.

Humans have been making star maps for thousands of years: there are star maps painted on the ceilings of

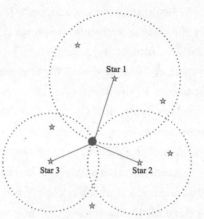

The determination of the coordinates of a planet using angular separations from three bright stars whose positions are well known. Each angular separation gives the radius of a circle around the reference star, and the three circles intersect at only one point.

Egyptian tombs that are accurate enough to date them to around 1500 BCE and Babylonian catalogues of stars dating to around 1200 BCE. In China, Su Song compiled an atlas of more than 1,400 stars around 1092 CE (based on earlier work dating back to the 600s CE), recording their positions on the great celestial globe driven by his tower clock. The most significant European stellar atlas was the catalogue of around 1,022 stars made by Claudius Ptolemy in the *Almagest*.‡

---

* A paper in the *American Journal of Physics* by Kevin Krisciunas describes this process in detail in the context of using very simple instruments to measure the orbit of Mars. This was published in October 2019, and thus arrived at a particularly fortuitous moment in the writing of this book.

† This idea of intersecting circles will be used again when we talk about GPS navigation in Chapter 13.

‡ Ptolemy's star catalogue was an update of the 850-star catalogue of Hipparchus from about 135 BCE, which was in turn building on earlier work by other Greek astronomers. The original text was lost after the fall of Rome and only preserved through translations in the Muslim caliphates from around 800 CE, which is why the book is now known by an Arabic-derived name.

We don't know what sort of star maps the Maya had, though there are hints in the codices and some paintings that they, too, divided the strip of sky where the sun, moon, and planets may be found in a set of constellations. The exact names of these, and how they map onto the European zodiac, are still a matter of debate in the archaeological community, and we'll probably never know for sure. It's likely, though, that they had some set of bright stars whose positions were well known and were able to use these as landmarks to track the changing motion of the planets in some detail.

Modern stellar atlases specify the positions of the stars in terms of two angular coordinates, much the same way we specify the positions of points on the surface of the earth using latitude and longitude. The celestial analogue of latitudes is called "declination," and the longitude analogue is "right ascension."* Just like latitude, declination is expressed as an angle away from the celestial equator, with the north celestial pole having a declination of +90 degrees and the south celestial pole a declination of –90 degrees. However, right ascension is expressed in units of *time*, with values traditionally recorded in hours, minutes, and seconds.†

Right ascension is connected to time because it's the angle in the direction of the earth's rotation, and thus constantly changing through the night. The right ascension value for a given star is closely related to the time elapsed between the 0h 0m 0s point being directly overhead and the star of interest being directly overhead:‡ if you precisely record the time of the "transit" when the star crosses the meridian, it's a simple matter to convert that to the right ascension value. For this reason, many older observatories are equipped with "transit telescopes," fixed to a mount that's strictly oriented along a north-south line. These can often be oriented to different declination values, but they will always be pointed exactly at the meridian, to catch the

---

* Many historical measurements, including most of those discussed here, were taken using latitude and longitude measured relative to the ecliptic, the path that the sun traces through the course of the year.

† Confusingly, angular separations measured in degrees are also subdivided into minutes (60 to the degree) and seconds (60 to the minute), but these are not the same size as the minutes and seconds of right ascension. One minute of right ascension is 15 minutes of arc.

‡ There's a slight correction needed to account for the difference between sidereal and solar hours; the sky turns through an hour of right ascension in very slightly less than an hour of conventional time.

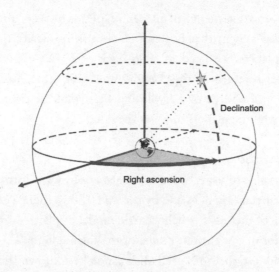

The position of a star seen from the earth is specified by two angular coordinates:
right ascension (analogous to longitude) and declination (analogous to latitude).

precise moment of transit, and establish either the precise time (if the right ascension of the transiting star is already known) or the right ascension of an unknown object.

We know very little about how the Maya measured time within the day, making it hard to say how they would've handled right ascension, but measuring *differences* in angles would have been well within their ability. All you need to make measurements of angles between two lights in the sky is an object held some distance from your eye so that it spans the distance between the two. The ratio of the length of the object to the distance from your eye is simply related to the angle. Images of Mayan astronomer-priests sometimes show them looking over pairs of crossed sticks, or holding a forked staff with a star symbol between the branches, suggesting that they did, in fact, have the tools needed for measuring and recording the fine details of planetary motion.

Whatever records the Maya may have had of planetary positions, and whatever models they may have used to predict the motion, have been lost to the ravages of time and weather or were consigned to the fire set by Diego de Landa and his compatriots centuries ago. They were almost certainly aware of the retrograde motion, and they had sufficient skill to be able to closely track it, but we will likely never know how well, or what model they used to make predictions.

To see how measurements of planetary positions were used to create a model of the solar system that brings a clockwork regularity to the changing motion of the planets, we need to cross back over the Atlantic. In the days when the Gregorian reform of the calendar was being set in motion, European astronomy was on the verge of a revolution that would settle once and for all the question of why the planets move as they do.

Orbits and Epicycles

In Europe, thousands of years' worth of observations went into the generation and testing of planetary models, and by the mid-1500s, there were two primary contenders. One of these is the heliocentric model of planets orbiting the sun, the direct ancestor of our current model of the solar system. This was a relative newcomer, introduced in 1543 by a Polish canon best known through the Latinized version of his name: Nicolaus Copernicus.* His most important work, *De revolutionibus orbium coelestium* ("On the revolutions of the celestial spheres"), was published when Copernicus was on his deathbed,† and while it made waves in scholarly circles, it was dense and difficult to follow. The heliocentric model gained in popularity thanks in part to the "Prutenic tables," a more user-friendly work published in 1551 offering predicted positions for all the planets calculated from Copernicus's model by the astronomer Erasmus Reinhold.

The more traditional alternative to the Prutenic tables was the Alfonsine tables, first compiled in the mid-1200s by a group of distinguished astronomers called together by Alfonso X, King of Castile. These were based on translations of Arabic tables assembled in Toledo, Spain, when it was under Muslim control, which in turn were based on still older compilations of data, tracing all the way back to Ptolemy's *Almagest* in Roman Egypt around 150 CE. A Latin version of the Alfonsine tables appeared in Paris in the early 1300s and rapidly spread through Europe. By 1500, they were the standard reference work for astronomers and astrologers across Europe. The Ptolemaic system used to calculate

---

* As always with people of this era, Copernicus's actual name is rendered in a dizzying variety of spellings depending on the exact source. In modern Polish, it would be rendered Mikołaj Kopernik.

† He had circulated a shorter draft version among friends and colleagues for many years before this, though. He was also drawing on ancient sources for inspiration, chiefly Aristarchus of Samos, who suggested the sun as the center of the universe around 250 BCE.

positions for the Alfonsine tables is geocentric: the moon, the sun, and each of the planets moves in a circular orbit around the stationary Earth.

Copernican and Ptolemaic models each offer their own conceptual explanation for the retrograde motion of the planets. In the Ptolemaic model, this is a direct addition to the motion of the object in question: each planet is actually moving in a small circular orbit called an "epicycle" around a point that is moving in a larger orbit around the stationary Earth. The occasional reversals that we see happen because the planet *really is* moving backward for a brief period.

The Copernican model, on the other hand, has the earth as yet another planet in orbit about the stationary sun, and explains the retrograde motion as a matter of different orbital speeds: retrograde motion happens when one planet overtakes another. To use the specific example of Mars, about once every two years the earth catches up with the slower-moving Mars, and for a moment the two planets lie on the same straight line passing out from the sun. During the period right around this moment of "opposition," the line of sight from the earth to Mars changes direction rapidly, leading to the illusion of Mars moving "backward" against the more distant stars.

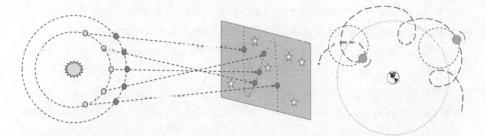

The retrograde motion of an outer planet as explained in the two different systems. On the left, the Copernican model depicts retrograde motion as the result of a fast-orbiting inner planet overtaking and passing a slower-orbiting outer planet, changing the viewing angle. On the right, the Ptolemaic model has the planet moving along an epicycle whose center orbits the earth, and thus at times the planet really is moving backward.

Both the Copernican and Ptolemaic models were thus able to explain the conceptual basis of retrograde motion in a relatively straightforward way. Predicting the exact positions of the planets seen in the sky was a more complicated process, but again, both models could manage it tolerably well, though both were far from perfect. Contrary to modern myth, the predictions of the Copernican system of the Prutenic tables were not dramatically better than those of the Ptolemaic system of the Alfonsine tables. Both sets of tables *claimed* an accuracy of around one-sixth of a degree in angle units, or about a third of the size of the full moon, whereas the actual predictions of the planetary models rarely came close to this level of accuracy and were often off by a degree or more. In the end, cracking the problem of the planets would require dramatically better measurements, which were finally provided through the work of a colorful Danish nobleman who may well have been the greatest naked-eye astronomer of all time.

A NOSE FOR ASTRONOMY

Tyge Ottesen Brahe, better known to history as "Tycho," the Latinized form of his name that he used as a scholar, was both brilliant and a bit arrogant. At the peak of his powers, the giant mural quadrant in his observatory was adorned with an enormous painting of himself surrounded by astronomical instruments, which he liked so much that it featured prominently in the frontispiece of the book he wrote describing his instruments. He also had a somewhat volatile temper: as a university student, he fought a duel with a distant cousin over some perceived slight, receiving a serious wound to the face that forced him to wear a prosthetic nose for the rest of his life.*

Tycho's career path was set in 1563 CE when, as a 16-year-old student in Leipzig, he observed a conjunction of the planets Jupiter and Saturn, a time when both planets had the same right ascension. While it was well established that a conjunction would occur, both the Prutenic and the Alfonsine tables were wrong about the date and location of the conjunction, with the Alfonsine tables missing the date by a full month.† Young Tycho decided that better astronomical tables were needed, and he was the man to produce them.

---

* Legend says his false nose was gold or silver, but chemical analysis of his exhumed body suggests that while he may have had precious-metal versions for special occasions, his daily wear fake nose was most likely brass.
† The Prutenic tables fared better for this specific conjunction, getting the date within a few days, but this was largely a matter of luck, as they were just as badly wrong about the dates of some other astronomical events.

There were strong social norms in Europe at the time against an aristo-crat of Tycho's stature—child of two of the most powerful noble families in Denmark—pursuing a scholarly career, but once he bulled his way past those, his status brought him great advantages. As a nobleman friendly with King Fred-erick II, Tycho was able to marshal resources unmatched by any other astrono-mer of his era. In 1576, he was granted the island of Hven in the Øresund strait as a fief and set about constructing a palatial estate/observatory there, which he dubbed Uraniborg, the "Castle of Urania," after the Greek muse of astronomy. This featured observational platforms and instruments incorporated into the design, along with the usual bedrooms and feasting hall. When Tycho decided he needed still larger instruments, a second observatory area, Stjerneborg ("Cas-tle of the Stars"), was constructed nearby. Between the rents he was due as lord of Hven* and the income from ceremonial positions elsewhere in Denmark, he had the funds to buy or build the very finest astronomical instruments.

Of course, fine instruments are of limited use without skilled hands and eyes to manipulate them, and Tycho was one of the best. Even before he was settled on Hven, he had begun to make a name for himself in scholarly circles, thanks to two important sets of observations that challenged Aristotelian ideas of how the world was organized.

In 1572, while living with his uncle in Herrevad Abbey, Tycho was sur-prised to spot a "star" in the vicinity of the constellation Cassiopeia that hadn't been there previously. This was, we know now, the result of a star several thou-sand light-years from the earth exploding as a supernova, but in Tycho's time there was no good explanation for a bright new object appearing so suddenly. The new star brightened rapidly, eventually becoming nearly as bright as Venus, then slowly faded from view over a period of a couple of years.

This event created a sensation among scholars across Europe and was observed by most of the notable astronomers of the day, but Tycho's obser-vations, eventually collected in a book, *De nova stella*, were among the best. He carefully measured the angular distance between the new star and several nearby stars to record its position, and rechecked these multiple times to see if

---

* Tycho did not endear himself to the residents of his remote island holding, who had been left more or less to their own devices for many years, and were none too happy when this new nobleman showed up and began commanding them to pay for and work on his various building projects.

they changed. The repeated position measurements let Tycho look for the star's parallax, an apparent change in position caused by a change in viewing angle.* With the transportation technology available in his time, Tycho wasn't able to quickly change his own location by any significant amount, but he didn't need to: all he had to do was wait and allow the rotation of the sky to do it for him. Through the course of the night, as the new star made its loop around the pole,† the viewing angle changed substantially. If the new star were relatively nearby, this should cause it to shift relative to the other stars. To the best of Tycho's ability to measure, though, the position of the new star did not change.

The lack of parallax for the new star showed that it was very distant from the earth, much more distant than the moon, which showed observable parallax. This was a clear problem for the Aristotelian worldview, which divided the universe into a variety of spherical zones associated with the various celestial objects located in them. The most distant sphere, containing the fixed stars, was held to be perfect and unchanging, with more transient effects like meteors and comets being essentially atmospheric phenomena confined to the "sublunar" region between the earth and the moon.‡ Tycho's parallax measurements placed the new star well outside that zone, and thus posed a philosophical problem.

The second significant astronomical event in Tycho's early career came while his great observatory was still under construction: the Great Comet of 1577 was visible around the globe. There are records of observations from Peru, China, and Vietnam as well as all over Europe and in the Islamic world. Again, Tycho used careful parallax measurements to determine that the comet was beyond the orbit of the moon; he was also one of the first to note that the tail always pointed away from the sun and that the speed of the comet slowed as it

---

* If an object is relatively nearby, it will appear to shift position against a more distant background when you change the observation point: a classic simple demonstration of this is to hold up a finger and look at it with first one eye and then the other. Your finger should appear to move between two different points relative to the background; the closer it is to your eyes, the greater the change in apparent position. This is why we have binocular vision: the difference between our eyes allows us to make a reasonable estimate of the distance to nearby objects, a great aid to moving around in a three-dimensional world.

† The constellation Cassiopeia is close enough to the celestial pole that at the latitude of Denmark it never sets.

‡ Thus the term "meteorology" for the study of weather.

moved away from the sun. He didn't publish his findings openly until a decade later,* but when he did, they helped cement his reputation as one of the greatest astronomers in Europe.

Tycho and the Planets

When his great observatory was fully operational, Tycho had the finest collection of astronomical instruments in all of Europe. Sadly, none of these have survived to the present day, but we know a great deal about them because of the aforementioned book he published bragging about them and describing their properties in considerable detail.

Frontispiece illustration from Tycho Brahe's book on his instruments, showing his giant mural quadrant with a painting of himself on it.

Astronomical instruments in Tycho's day measured angles using the same basic principle as aiming a gun with fixed sights: you define a line of sight by moving around so a distant marker is aligned with the target object. With that line established, you put a second marker on a second object—the use of two sight lines is what clearly distinguishes astronomers from teens shooting zombies in a video game[†]—and then measure the angle between the lines from your eye to each of the two markers. This procedure turns an angle measurement into a distance measurement: you measure the distance between the markers and the distance from your eye to the markers, and do a little

---

[*] His observational notebook from that time survives and provides an interesting glimpse into his thought processes, but the only immediate product was a report to the king that was largely concerned with the astrological significance of the comet for political events in the coming years. This sort of reporting on signs and portents was part of his duties for the crown, and fairly typical for the day.

[†] Though not characters in a John Woo movie.

trigonometry to convert that to an angle.* The greater the distance from your eye to the markers, the smaller the difference in angles you can measure—this is part of why a rifle is more accurate than a pistol—so if you want to do really accurate measurements, you need really big instruments. Tycho's observatory pushed the limits of both size and quality.

His most famous instrument was the huge mural quadrant, essentially a circular arc a bit under two meters in radius painted on a north-south wall with markings for angles from 0 to 90 degrees (one-fourth of a circle, thus "quadrant"). The observer would move around to sight the object of interest as it crossed the north-south meridian, reading the declination off directly and determining the right ascension from the time (measured by water clocks in the observatory; in Tycho's heyday, mechanical clocks weren't up to the level of precision he demanded).

For greater flexibility, he also had rotatable versions of the quadrant, also a couple of meters in radius, that could be reoriented away from the north-south line of the great mural quadrant; the first of these was built of wood, later replaced with a sturdier steel version. On a slightly smaller scale, he had brass quadrants that could be tilted to any angle, and later in his time at Hven, sextants operating on the same principle, but with an arc of one-sixth of a circle (60 degrees).†

Tycho's best angular measurements were accurate to within a few minutes of arc, about one-tenth the size of the full moon. He was thus well positioned to make a definitive determination of how the solar system was organized, through a combination of improved measurements of the positions of reference stars as well as better measurements of the planetary positions. With a characteristic lack of humility, Tycho decided that both the Ptolemaic and Copernican models of the solar system were deficient. He preferred a hybrid model of his own devising, in which the other planets orbit the sun, while the sun and the moon orbit the earth. This "Tychonic" system is geometrically the same as the

---

* Or you do a bit of trigonometry when you're *making* the instrument, and indicate the angle associated with various marker positions; either way, there's no escaping some math.

† Sextants were somewhat lighter than quadrants, and thus less prone to bending under their own weight.

Copernican, but it was more aesthetically appealing to Tycho, as it allowed for the dense and heavy Earth to be at rest, while the various lights in the sky were in constant motion. They would've needed to be in extremely rapid motion, but at the time nobody had any idea how enormously massive the sun and planets are, so that was not necessarily a flaw.*

All of the world systems being promoted in Tycho's day were able to explain the basic phenomena of planetary motion, making it difficult to distinguish between them, but there was one clear difference having to do with the planet Mars. In the Ptolemaic system, where all bodies orbit the earth, Mars is always more distant from the earth than the sun. In both the Tychonic and Copernican systems, however, Mars orbits the sun, and at the right times of year is closer to the earth than to the sun. Tycho hoped he could use the technique that had served him well with both the new star and the comet, and measure the parallax of Mars to determine its distance from the earth. He made the parallax measurement a key focus of his observational efforts for much of the 1580s, spending significant time observing Mars during 1982, 1983, 1985, and 1987, when Mars was close to the earth during the winter months, and the long Danish nights allowed the maximum possible time between observations of the red planet. Tycho and his assistants would record the angular separations between Mars and selected reference stars in early evening, as Mars crossed the meridian, and again just before sunrise, hoping to find a shift in the position.

In 1587, Tycho briefly believed that he had measured the parallax of Mars, but a careful reanalysis two years later showed that his measurements were hopelessly distorted by the refraction of light in the earth's atmosphere. When stars or planets are closer to the horizon, their light must pass through a much greater thickness of air than when they're near the zenith, which leads to some deflection of the light, in the same way that the changing thickness of glass in a lens does. Tycho's 1587 measurement used an older value to correct for this refraction effect, but a more careful study of refraction obliterated the parallax

---

* Another astronomer, Nicolaus Reimers Bär, began promoting a similar system in the late 1580s, albeit with a rotating Earth. Tycho believed Bär had plagiarized his idea during a visit to Uraniborg in 1584, and as a result dedicated significant effort to hounding Bär in various courts and getting Bär's book banned throughout the Holy Roman Empire.

he thought he had seen. Tycho sadly abandoned his attempts to directly measure the distance to Mars at around this time.*

With a decisive parallax measurement out of reach, Tycho's next best hope
was that his exceptionally precise measurements of planetary positions could
allow the orbits to be determined well enough to find some other feature that
would definitively demonstrate the superiority of the Tychonic system. Tycho's
data were far and away the best in Europe, so he was in the best position to
settle this observationally, but the calculations to turn position measurements
into orbits were highly demanding and would require the full attention of an
expert mathematician. At around this time, though, Tycho fell out of favor with
the young King Christian IV of Denmark, who took the throne after his father's
death in 1588; Tycho became distracted with political concerns that eventually
forced him into exile in Bohemia.

Tycho's project to settle the structure of the universe was put on hold for a
time as he relocated, but while he did eventually establish a new observatory, he
never again reached the same heights as during his years on Hven. Tycho died
in 1601, his great work unfinished, but he left behind a collection of planetary
observations of unparalleled quality, in the hands of a brilliant young assistant
he met on his journey to Prague.

Both Tycho Brahe and the late Mayan astronomer-priests who drew up the
Dresden Codex were ultimately engaged in the same project: imposing a com-
prehensible order on the complex motion of the planets. They even had something of the same goal in mind: in Tycho's day there was not the sharp distinction
between astronomy and astrology that we have today, so his models also served
a divinatory purpose. Better predictions of the positions of the planets would

---

* Using what we know now about the size and arrangement of the solar system, Tycho
  never stood a chance of success: the maximum possible value of the diurnal parallax he
  was attempting to measure is less than half a minute of arc, several times smaller than the
  uncertainty of the best measurements his instruments could achieve. Nearly a century
  later, a Mars parallax shift of about 15 minutes of arc was finally measured by combining
  observations by Jean Richer's expedition to Cayenne (discussed in Chapter 6) with
  simultaneous observations several thousand miles away in Paris.

allow better astrological forecasts, and the casting of horoscopes was one of the chief duties of Tycho's roles as court astronomer in Denmark and Bohemia. In this respect, he was not so different from the Mayan priests compiling lists of (mostly bad) omens associated with the cycles of Venus.

Brahe and the Maya also represent complementary approaches to the general problem of astronomical prediction. The Maya drew on relatively simple observations over centuries, as seen in the occasional corrections directed in the Dresden Codex tables. Tycho, with access to dramatically better metallurgy and manufacturing, made a smaller set of observations over just a few decades, but they were of such exceptional precision that they would enable Tycho's last assistant and scientific heir, Johannes Kepler, to trigger a philosophical revolution in our understanding of the universe, as we'll see in the next chapter.

Chapter 8

CELESTIAL CLOCKWORK

One of the most influential scientific statements of the twentieth century was a postulate introduced in 1905 by a young scholar who had failed to secure an academic position and found himself working as a patent clerk in Switzerland. In a paper with the decidedly uninspiring title "On the Electrodynamics of Moving Bodies," written in dry academic prose, he asserted:

> *Any ray of light moves in the "stationary" system of co-ordinates with the determined velocity c, whether the ray be emitted by a stationary or by a moving body.* *

Translated into more colloquial language, this says that the speed of light is the same to all observers, regardless of how the source may be moving: if you measure the speed of the light emitted by an electron accelerated to 90 percent the speed of light in a synchrotron and compare that to the speed of the light emitted by the stationary monitor on the computer you used to collect the data, you will find they are exactly the same.

---

* Strictly speaking, what he actually said was "*Jeder Lichtstrahl bewegt sich im 'ruhenden' Koordinatensystem mit, der bestimmten Geschwindigkeit V, unabhängig davon, ob dieser Lichtstrahl von einem ruhenden oder bewegten Körper emittiert ist.*" because the paper in question was written in German, and published in the *Annalen der Physik*.

The paper containing this remarkable assertion is the first introduction of Albert Einstein's theory of special relativity, one of the founding documents of modern physics. The idea of the speed of light as a universal constant has far-reaching consequences for our understanding of space and time that we will soon discuss at greater length.

While elevating it to a fundamental principle was new, the idea that light has a finite speed was a well-established empirical fact in 1905, and had been for over 200 years. Einstein didn't give a numerical value for the speed of light in his paper; by that time the best measurements were within 1 percent of the modern value of 299,792,458 meters per second. Such a high speed is extraordinarily difficult to measure, leading some early philosophers to speculate that it might be infinite, but it was conclusively shown to be finite (and its value measured) by the Danish astronomer Ole Christensen Rømer all the way back in 1676.

Rømer's measurement was based on observations of one of Jupiter's moons, Io, specifically a difference of a few minutes between the times when it would disappear behind the planet and the times when it was *expected* to disappear. This was made possible by two revolutionary developments in technology: the invention of the pendulum clock—and the associated leap in timekeeping precision—and the invention of the telescope, which enabled the discovery of Jupiter's four moons. Just as important was the philosophical revolution that produced a reliable mathematical model for predicting the positions of bodies in the solar system. The precise observations Rømer made would not have been sufficient by themselves without the new understanding of the solar system and its organization that emerged in the early 1600s. This was built on the observations of Tycho Brahe by one of his final assistants, the brilliant Austrian mathematician Johannes Kepler.

KEPLER'S OBSESSIONS

Where Tycho Brahe was born to wealth and privilege, Johannes Kepler, 25 years his junior, was the child of a once prosperous family fallen on hard times. His father made a poor living as a mercenary and was an intermittent and violent presence in the household before disappearing entirely when Kepler was 16 years old. As a child, Kepler was weak and sickly, and a near-fatal case of small-pox left him with poor vision. He had an undeniable talent for mathematics, though, and performed well in school when the family finances allowed for him to attend. His teachers recognized his potential and helped him to attend seminaries at Adelberg and Maulbronn, Germany, then eventually the University of Tübingen. Some of the university faculty presciently wrote, in a letter recommending him for a scholarship, that "Young Kepler has such an extraordinary and splendid intellect that something special can be expected from him."

After completing his university studies, Kepler was assigned as a teacher of mathematics and astronomy in a school in Graz, Austria. The position was reasonably well suited to his talents, but the location made his situation precarious: Kepler was an ardently religious Protestant,* but Graz was in a Catholic area, and Kepler and his family were in constant danger of being forcibly converted, banished, or killed.

While teaching in Graz, Kepler carried on research in astronomy, leading to the publication of his first book, *Mysterium Cosmographicum*, in 1596. His writings stemmed from an epiphany he had regarding the six known planets and the five Platonic solids. Kepler argued for a Copernican system with orbits determined by these solids: the orbit of Mercury lies on a sphere that just fits inside an octahedron that is bounded by the sphere containing the orbit of Venus. The orbit of Venus is contained within an icosahedron that is bounded by the sphere containing the orbit of the earth. Earth's orbit is contained within a dodecahedron, then Mars's orbit fits inside a tetrahedron, and finally Jupiter's orbit fits inside a cube within the orbit of Saturn.

To modern eyes, this seems like a great deal of mystical claptrap, but as noted, the lines between astronomy, astrology, and religion were not as clearly drawn in the late 1500s as today. Kepler was firmly convinced that God must have arranged the universe in a pleasing and harmonious manner, and this obsession with harmony guided all his explorations. In Kepler's mind, this Platonic-solid arrangement of orbits was just such an expression of profound and elegant harmony within creation. More importantly, the arrangement of

---

* He was a Lutheran by upbringing and practice, but occasionally got in trouble even with his own sect for being willing to seriously consider the arguments of Calvinist preachers. This intellectual open-mindedness would serve him well in science, but it did not make for an easy life navigating the religious schisms of Europe around 1600.

shapes agreed reasonably well with then current estimates of the sizes of the orbits,* and it provided a relatively simple rule for determining the orbital periods of the planets in terms of the radii of the spheres.†

Kepler sent copies of his book to many prominent scholars in hopes of finding a way out of Graz. Among these was Tycho Brahe, who was by this time making his way slowly to Prague, seeking to associate himself with the court of the Holy Roman Emperor Rudolf II. While he did not accept the Copernican model of Kepler's book, Brahe recognized the mathematical talent required to write it, and in 1600 invited Kepler to visit him. The invitation could not have come at a better time for Kepler and his family, who were on the verge of being officially banished from Graz for their religious beliefs.

Tycho hired Kepler as an assistant and put him to work on the calculations needed to convert his years of measurements of the position of Mars into an orbit for the red planet. Their working relationship was a stormy one, as Tycho was very secretive and strictly limited Kepler's access to his data. Kepler himself was, in modern parlance, somewhat drama-prone, so in the time they worked together, they had multiple blowups and reconciliations. Kepler was unquestionably one of the most talented mathematicians available, though, so he gradually earned Tycho's trust and gained greater freedom and access to the data. In the early fall of 1601, Tycho secured a commission from Emperor Rudolf for a grand project, a new set of astronomical tables based on Tycho's observations, with computations to be handled by Kepler.

Unfortunately, the collaboration between Kepler and Brahe was short-lived. Just days after getting imperial approval for the new project, Tycho fell ill and died on October 24, 1601.‡ His last words, recorded by Kepler, were "Let me not seem to have lived in vain."

---

* Strictly speaking, all that was known was the ratio of the various orbits to the orbit of the earth; the absolute scale of the solar system wasn't nailed down until many years later.

† He wasn't satisfied with this, but with some additional work, he discovered another relationship that he found philosophically captivating: ratios of the orbital periods of the planets seemed to match those in music, leading Kepler to think of the solar system as playing a sort of heavenly chord. Again, this seems weird to modern minds but was not out of line for a serious scholar in Kepler's day.

‡ Tycho suffered a ruptured bladder while a guest at a banquet: it would have been a breach of etiquette to get up before the host, even to relieve oneself, so Tycho remained at the table through what must have been extreme discomfort. This story is famous—and struck many as

Kepler's Laws of Planetary Motion

The commission for the new tables continued past Tycho Brahe's death, with Kepler assuming full responsibility and receiving the title "Imperial Mathematician." The process of producing the tables was not smooth, though, and was held up by years of legal wrangling with Tycho's heirs, Rudolf being deposed as emperor and replaced by his younger brother Matthias, and the start of the Thirty Years' War. Also, more relevant to the subject of this book, Kepler needed to develop an entirely new set of principles to understand the organization of the solar system and make it as predictable as clockwork.

Armed with Tycho's observations of Mars covering the better part of 20 years, Kepler began work on determining its orbit with greater precision than any past astronomer would've been able to manage. While Tycho had, even on his deathbed, urged Kepler to promote the Tychonic system, Kepler worked within the Copernican paradigm that he preferred, putting the sun at the center of the orbiting planets, including the earth. The process of determining orbits is a tedious one, involving many repeated calculations in trigonometry to determine the radius of a planetary orbit in terms of the radius of the earth's orbit. While Kepler maintained his position as Imperial Mathematician throughout, that was not a lavishly compensated position, and Rudolf's habit of overcommitting resources meant that delivery of his salary was somewhat erratic. As a result, Kepler was unable to hire an assistant and did all the calculations himself. In his *Astronomia Nova* published in 1609, he wrote: "If thou art bored with this wearisome method of calculation, take pity on me, who had to go through with at least seventy repetitions of it." When the time came to present his work to the world, Kepler delivered an argument that to a large extent paralleled the process by which he discovered the new system. To ease his readers into the new system, he began by using apparatus familiar to both Copernican and Ptolemaic astronomers.

Astronomers of both persuasions had long realized that they could not reproduce observations of planetary motion using perfectly circular orbits centered on either the sun or the earth. The orbits and observations fit together

---

too good to be true, to such a degree that Brahe's body was exhumed in both 1901 and again in 2010 to investigate other explanations (kidney stones or mercury poisoning). To the best anyone can determine at this remove, the famous story appears to hold up.

better if they shifted the center of the orbit to an "eccentric" point some distance from the central body. This only partially accounted for the tendency of planets to move faster at some times than others, so the orbital speed still needed to change as the planets moved around the orbits. Ptolemy's great innovation was a systematic way to calculate this: he introduced an "equant" point on the opposite side of the eccentric, from which the planet in question appears to move at a constant angular speed. Mars moves a little slower when it's on the equant side of the orbit, and a little faster when it's on the earth/sun side, in such a way that an imaginary observer standing at the equant would see it always shift its longitude by the same amount in the same direction.

Kepler's attack on the orbit of Mars was a kind of combination of the two—putting the sun at the center in the Copernican manner but using an equant point like a Ptolemeian—but the techniques for working with these were well established. He then used Tycho's observations to determine the relative locations of the eccentric and the equant by two different methods.

One approach started with a handful of observations Tycho had made when Mars was in opposition—that is, when the sun, the earth, and Mars all fell along a straight line. In these cases, the "longitude" of Mars seen from the sun is unambiguous; Kepler used these data to construct a model to predict this longitude and found that it matched eight other such observations perfectly, usually to within about one minute of arc, less than the uncertainty in Tycho's observations. For the same four points he started with, he also calculated the longitude as seen from the equant point. In this case, he knew the elapsed time from the dates of the observations, and since by definition Mars moves at a constant speed when viewed from the equant, it's a simple matter to convert this to changes in angular position.

This method of calculation gives two sets of four longitudes, starting at two different locations, pairs of which must end up pointing at the actual position of Mars in its orbit. There is one and only one way to arrange the equant and eccentric so that the four Mars positions fall on the same circular orbit. Taking the sun as fixed, Kepler adjusted the location of the equant to find the point where these four angles coincided with the positions of Mars, and then found the center of the circle connecting those positions. This gave him the locations of the eccentric and the equant relative to the sun; the eccentric found by this process ends up a little closer to the equant than would have been expected from the traditional Ptolemaic approach, which would put it exactly at the midpoint between the sun and the equant. But Tycho's observations were of such high quality that there could be no doubt regarding the difference.

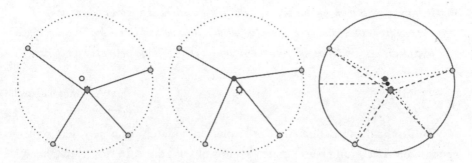

Kepler's method for determining the orbit of Mars: First, calculate the longitude angles of four Mars positions as seen from the sun. Second, find the longitude angles of those same positions as seen from the equant. Finally, combine the two sets of angles to determine the center of the orbit that includes all four points.

Kepler had another way to measure the location of the eccentric, though, this one using the "latitude" of Mars. The orbit of Mars is inclined slightly relative to that of the earth, which causes it to move up and down in declination, oscillating between two extremes. Tycho's data also allowed Kepler to fix the location of those extreme points, and determine the center of Mars's orbit that way. When he did that, the eccentric point ended up almost exactly midway between the sun and the equant.

So, Kepler now had two different methods for locating the center of the circular orbit for Mars, which gave two different answers. If he used the eccentric position measured by the longitudes, the distances of the extreme points in latitude came out badly wrong. On the other hand, if he put the eccentric where Ptolemaic thinking and the latitude measurement would have it, the predicted longitudes came out wrong.

The absolute difference in predicted and observed longitudes is quite small, around eight minutes of arc, or a quarter the width of the full moon. This would've been deemed a resounding success in an earlier era—Ptolemy claimed an accuracy of only plus or minus 10 minutes for his observations. An eight-minute discrepancy from the observations of *Tycho Brahe*, though, was not an error that could be ignored. As Kepler put it:

> *Therefore, something among those things we have assumed must be false. But what was assumed was: that the orbit upon which the planet moves is a perfect circle; and that there exists some unique point on the line of apsides at a fixed and constant distance from the center of the eccentric about which point Mars describes equal angles in equal times. Therefore, of these, one or the other or perhaps both are false, for the observations used are not false.*

> —*from* Astronomia Nova

This eight-minute difference set Kepler on a search for a new heliocentric system, one that would better match Tycho's observations. In this search, he was also guided by his philosophical convictions. He had become convinced that attention must be paid to what *caused* the motion of the planets, finding the idea of orbits centered on otherwise empty points in space distasteful. And, of course, he had his lifelong conviction that God would have ordered the world in a way that had elegance and harmony.

Having rejected eccentric circular orbits, Kepler began to consider a variety of oval orbits, but ran into the problem that "oval" as a mathematical category is not perfectly well-defined. There are a great many oval shapes that might work but can't be conveniently described in a mathematical form. The one oval that *does* have a precise and convenient mathematical definition is an ellipse: where a circle has a single center point and every point on the circle has the same distance from the center, an ellipse has two focal points, and every point on the ellipse has the same *sum* of the distances to the focal points.* Kepler initially rejected the idea of elliptical orbits, thinking that they were too obvious—if the answer were as simple as replacing circles with ellipses, surely some earlier astronomer would've figured that out. For mathematical convenience, though, he began using an ellipse to approximate a more general oval, and then stumbled on the true answer: the orbit of Mars was an ellipse, with the sun at one of the focal points.

---

* This leads to the simple trick for drawing an ellipse, namely putting a loose loop of string around two nails in the positions of the foci. A pen placed inside the loop so it stretches the string taut will trace an ellipse when dragged all the way around the nails.

This solution addressed all of Kepler's needs: it's a shape with a precise mathematical description, surely appealing to the benevolent God who created mathematicians, and placing the sun at one of the two focal points allows for the orbits to be determined by some force emanating from the sun. It also fit with an idea he had been working on for some time: that the speed of a planet in its orbit varies in a way that depends only on its distance from the sun. Planets move faster as they approach the sun, and slower as they recede from it, and Kepler found a geometrical formula for the relationship: a line between the planet and the sun sweeps out equal areas in equal times.*

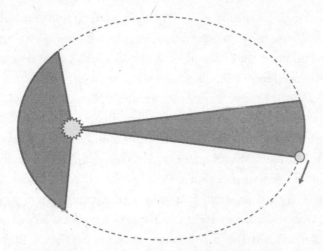

According to Kepler's second law, the areas of the two shaded regions
are the same, and the planet moves through the corresponding
sections of its elliptical orbit in the same amount of time.

Kepler's *Astronomia Nova* was a powerful argument for his modified Copernican system based on Tycho's observations. He explained the inadequacy of the eccentric-and-equant system for describing the orbit of Mars (as discussed above) and also argued that the earth was a planet, using an ingenious scheme

---

* This "area law" eventually became known as Kepler's second law of planetary motion, even though it was the first one he actually discovered; the first law (discovered second) is that the orbits of the planets are ellipses with the sun at one focus. There's also a third law relating the period to the size of the orbital ellipse, which Kepler didn't formulate until his *Harmonices Mundi* in 1619.

where he reversed the Mars observations to determine the properties of the earth's orbit. Finally, he detailed his area law and discovery of elliptical orbits.

This was followed by the *Epitome Astronomiae Copernicanae* between 1617 and 1621, a textbook on how to perform computations within his system, and the *Hamonices Mundi* in 1619, a larger discussion of "harmony" in a geometrical sense, which ends with his third law of planetary motion describing the orbital periods of the planets. Finally, in 1627, after years of legal wrangling and delays caused by political strife, he published the completed *Rudolphine Tables* (by this time, two emperors past the man who gave him the job).

While Kepler's speculations about the sun as the source of planetary motion were less than enthusiastically received, there was no arguing with the accuracy of his tables. Where the previous tables regularly had errors of a degree or more in the predicted positions of the planets, Kepler's tables were generally good to within several minutes.* The tables, and the underlying elliptical system, were accurate enough to predict transits of the inner planets, moments when they pass directly between the earth and the sun. These transits are an incredibly stringent test because the sun is only about 30 minutes of arc wide, and transits typically last just hours, but Kepler's system passed this test with flying colors. Pierre Gassendi used a prediction from Kepler's tables to observe a transit of Mercury in 1631, and Jeremiah Horrocks used Kepler's system to predict and observe a transit of Venus in 1639 CE. These observations helped seal the case for the Keplerian model of a heliocentric solar system with elliptical orbits, which became generally accepted by the middle of the 1600s.

Kepler himself did not live to see these transits, alas. He died in November of 1630, falling ill after a cold and miserable journey as part of yet another attempt to collect back pay he was owed, having been chased from yet another city by the Thirty Years' War.† His scientific immortality was ensured, though, as his new brand of Copernicanism was spreading rapidly, with his book becoming the standard text. He had even played a key part in establishing the next

---

* They weren't perfect, though; the calculations were complicated and tedious, and the printing involved hand-set type, so inevitably errors crept in. The underlying system allowed vastly improved accuracy, though, particularly for observers with the mathematical skill to do the calculations themselves.

† These were two sadly recurring themes in his life; he also spent several years defending his elderly mother when jealous neighbors accused her of witchcraft.

age of astronomy with the publication in 1611 of *Dioptrice*, the first detailed treatise on the optics of the telescope.

THE MOONS OF JUPITER

In late September 1608, a Dutch spectacle maker named Hans Lipperhey presented a new invention to Prince Maurits of Nassau and a collection of other dignitaries assembled in The Hague for a peace conference. Lipperhey's device consisted of a hollow tube with lenses at either end: a simple telescope. A widely circulated French account of the demonstration noted the potential uses of the device:

> *The said glasses are very useful at sieges & in similar affairs, because one can distinguish from a mile's distance & beyond several objects very well, as if they are very near & even the stars which normally are not visible for us, because of the scanty proportion and feeble sight of our eyes, can be seen with this instrument.* *

News of the invention spread rapidly across Europe,[†] particularly in scholarly circles, and astronomers and natural philosophers quickly began constructing telescopes of their own (with varying degrees of success, as the technology and materials available for making lenses were quite poor). Among them was a fairly obscure professor of mathematics in Padua by the name of Galileo Galilei.

As we saw in Chapter 6, Galileo was a careful and clever observer whose ingenuity in devising experiments was exceeded only by his talents as a

---

* From *Ambassades du Roy de Siam envoyé à l'Excellence du Prince Maurice, arrive a La Haye, le 10. septembr. 1608* ("Embassy of the King of Siam sent to his Excellence Prince Maurice, September 10, 1608"). The telescope demonstration happened to coincide with a visit from a Siamese diplomat, so was mentioned in passing, then went on to become the most famous part of the report.

† Also the inevitable set of competing claims to be the true inventor of the telescope. While there are stories of other spectacle makers also offering such instruments at around the same time, Lipperhey's demonstration in The Hague is the first clearly documented account of a telescope in use.

self-promoter. Having learned of the telescope, and made one of his own,* Galileo recognized an opportunity for advancement, and he quickly arranged a demonstration for the aristocrats of Venice, formally presenting them with a telescope in August of 1609. For this, he was rewarded with a lifetime appointment, including an enormous salary, at the university there.

Galileo really struck gold, though, when he turned the telescope upward and began using it to make astronomical observations. He was not the first to do this—that honor probably goes to the English astronomer Thomas Harriot, who made a sketch of the moon as seen through his telescope in July 1609—but Galileo had the keenest grasp of the possibilities for career advancement from new observations. In January of 1610, he pointed his telescope at Jupiter and noticed that the bright disk of the planet was accompanied by several small bright lights. These changed position on subsequent nights, but always remained near Jupiter, which Galileo recognized as a momentous discovery: Jupiter had satellites of its own.

Galileo wasn't the only scholar to notice this in 1610—Simon Marius, the court astronomer of Ansbach (now part of Germany), first observed them one day later†—but he recognized the significance immediately. His observing notes switch from the usual vernacular to scholarly Latin, so as to be more quickly converted to a book, and a short time later he reached out to the powerful Medici in Florence to offer them the opportunity to name the new satellites. In March of 1610, Galileo rushed out a book, *Sidereus Nuncius*, detailing his telescopic observations and announcing to the world the "Medicean Stars," which we now know are the four largest moons of Jupiter. Galileo was promoted to the official mathematician and philosopher of the Medici court, and overnight he became one of the most famous scholars and astronomers in Europe.

The four moons discovered by Galileo now go by names drawn from Greek mythology, suggested by Kepler and first published by Simon Marius:

---

* Galileo did not invent or even substantially revise the design of the telescope, but he was a skilled instrument maker and produced telescopes of higher quality than many of the others being thrown together across Europe.

† Marius, being more careful and less ambitious than Galileo, didn't publish his observations until 1614, at which point Galileo viciously attacked him for supposed plagiarism.

Io, Europa, Ganymede, and Callisto (in order from closest to farthest).* These were a significant boon to the Copernican and Tychonic models of the solar system, as they provided proof that astronomical objects could orbit bodies other than the earth. They also provided another test bed for Kepler's laws of orbital motion: careful measurements of the sizes and periods of the orbits confirmed that they also fit Kepler's third law.[†]

Photo of Jupiter and the four Galilean moons, taken by Talha Zia.

Seen from the earth through a telescope, the Galilean moons of Jupiter appear as bright dots moving back and forth across the planet, with the four different moons reaching different maximum distances corresponding to the radii of their orbits. The periods of these orbits are fairly short, ranging from 1.8 days for Io to 16.7 days for Callisto, so their arrangement as seen from the earth changes rapidly. Once per orbit, each moon also slips into Jupiter's shadow and disappears from view for a couple of hours. If these orbits could be predicted

---

* These are, respectively: a mortal princess Zeus lusted after and turned into a heifer to hide her from the wrath of Hera; a Phoenician princess whom Zeus abducted in the form of a bull, who eventually became queen of Crete and the mother of King Minos; a Trojan prince of exceptional beauty abducted by Zeus in the form of an eagle and brought to Olympus to serve as a cupbearer; and a nymph in the service of Artemis whom Hera transformed into a bear after Zeus seduced her, who was eventually transformed into the constellation Ursa Major. The love lives of the Greek gods are really something.

[†] Measuring the orbital periods and comparing to Kepler's third law is now a common lab activity in introductory astronomy courses.

sufficiently well, the pattern of their positions and the timing of their eclipses could be used as a universal clock.

LIGHT SPEED

Galileo recognized the potential for using Jupiter's moons as a clock, but he never actually tracked the orbits well enough to draw up the necessary tables. That task was finally accomplished in 1668 by Jean-Dominique Cassini,* whose tables of eclipse timings were sufficiently precise to use for longitude measurements on land (we'll discuss the connection between time and longitude more in the next chapter). Cassini was the director of the Paris Observatory when Jean Richer was sent to Cayenne in 1672; when Richer got there, he set his clocks by observing eclipses of the moons of Jupiter. This clockwork regularity was the key to establishing the finite speed of light.

Part of the research program undertaken by the observatory in Paris included improved techniques for measuring the latitude and longitude of points on the earth, which is crucial information when converting astronomical observations into celestial coordinates. With new and improved measurements of their own location made, the French needed a similar upgrade in the precision of the recorded position of Tycho Brahe's observatory to merge his data with theirs, so in 1671 (70 years after Tycho's death), Cassini's colleague Jean Picard led an expedition to Hven to resurvey the location of Uraniborg. While there, he was sufficiently impressed with one of the local assistants, Ole Christensen Rømer, that he offered the young man a position in Paris.

Rømer was the son of a merchant seaman. From a very young age, he showed a wide-ranging curiosity and a gift for working with mechanical devices. He was educated in Aarhus and at the University of Copenhagen, where he worked with one of his professors on preparing a new edition of Tycho's observations. He jumped at the chance to work with Picard and Cassini in Paris, and spent nine years there on a wide range of projects, from attempting to measure

---

* This is the same Cassini we met in Chapter 6, at that time still working in his native Italy. In 1671, Louis XIV invited him to Paris to head up a new observatory there, and he became thoroughly naturalized, changing his name from Giovanni Domenico Cassini.

the parallax of stars to assisting with the design and construction of the magnificent fountains at Louis XIV's palace in Versailles.

While working on Cassini's tables of the orbits of the Galilean moons, Rømer noticed a pattern in the timing of the eclipses of Io, the innermost of the four.* While the average period of Io's orbit calculated from a full year of data (i.e., a bit more than 200 orbits) was very accurate and reliable, there were periods in the year during which the next eclipse consistently occurred slightly earlier than expected from the average period, while at other times each new eclipse came slightly later than expected. The time difference was small, but over a period of months it would add up to a few minutes, a difference easily detected using the high-quality mechanical clocks that had recently become available.

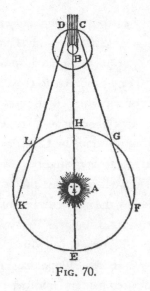

Fig. 70.

According to Rømer's method, eclipses that occur while the earth is moving from F to G happen earlier than predicted, while those that occur while the earth moves from L to K happen later.

Rømer realized that the too-early eclipses came during the months when the earth was approaching Jupiter, while the too-late ones showed up while the earth was moving away from Jupiter. This discrepancy points to light having a finite speed: it takes some time for light to cover the distance between Io and the earth, and changes in that distance change the time between eclipses.

The time from the start of one eclipse of Io to the start of the next is a bit less than two days, and during that time, the earth moves about 4.5 million kilometers in its orbit. If that motion is directed perpendicular to the line between the earth and Jupiter, it doesn't change anything, but if that motion is directed *toward* Jupiter, when the second eclipse comes around, the light has a shorter distance to travel and arrives about 15 seconds earlier than expected. If

---

* This works best for Io because it's the innermost moon, and thus its orbit is almost entirely determined by Jupiter. The orbits of the outer moons are perturbed by their mutual attraction, and thus a bit more complicated.

the earth is moving *away* from Jupiter, the light from the second eclipse has a longer distance to travel and arrives about 15 seconds later than expected.

Rømer estimated from his measurements of the delay that light would take 10 or 11 minutes to cross a distance equal to the radius of the earth's orbit; the modern value is 8 minutes, 19 seconds. As the actual size of the earth's orbit was not well known at the time, Rømer never estimated the speed in absolute units; that was done by Christiaan Huygens (designer of the first working pendulum clock, from Chapter 6), using Rømer's data on Io and some other observations. Huygens estimated the speed of light at 212,000 kilometers per second, well below the modern value of 299,792.458 km/s, but very respectable for the first such measurement.

Rømer returned to Denmark in 1681 as a professor of astronomy at the University of Copenhagen, where he presided over the observatory there and devised new and improved astronomical instruments; he also studied thermometry, devising an important forerunner of the Fahrenheit temperature scale. He held administrative positions both at the university and in the city of Copenhagen, where he was chief of police for a time, reformed the public works, and, somewhat ironically for an astronomer, established the first system of streetlights in the city. His most enduring contribution to science, though, remains his discovery that light moves at a finite speed, 220 years before Einstein established that speed as a universal constant.

Ole Rømer's discovery brought together three of the great scientific revolutions of the 1500s–1600s, two technological and one philosophical. He obviously needed the invention of the telescope in order to be able to see Jupiter's moons, and the development of pendulum clocks greatly simplified the process of making reliable astronomical measurements. Those would not have been enough, though, without Kepler's philosophical revolution, using Tycho Brahe's data. The regular and reliable elliptical orbits that Kepler established, based on a discrepancy of a mere eight minutes of arc, brought astronomy to the level of precision where Rømer could confidently attribute a difference of a few minutes in the time of an eclipse to a real physical effect, and make an important leap in our understanding of the behavior of light.

Chapter 9

TO THE MOON . . .

L ouis XIV is said to have joked that France had lost more territory to the resurveying of its borders by Jean-Félix Picard and astronomers from Jean-Dominique Cassini's Paris Observatory than it gained by military conquests. Like most jokes, it contains an element of truth: Improved time-keeping and the ability to find the time from the moons of Jupiter led to a dramatic improvement in the ability to determine the position of a point on the earth. By the early 1700s, clocks, telescopes, and astronomical tables were being put to use by surveyors to better fix the location of political boundaries. One important caveat, though: the new technologies did not work well on the deck of a ship at sea. As mentioned in Chapter 6, the rocking of the waves perturbed the motion of a pendulum clock too much to allow reliable oper-ation, and those same circumstances made it too difficult to keep a telescope trained on Jupiter well enough to observe eclipses of Io. This left a major gap in the ability of merchant and military seamen to navigate in the open ocean, something that became an enormous and expensive problem with the growth of globe-spanning empires through the 1600s.

Toward the end of Louis's 72-year reign, certainly by the early 1700s, the poor state of navigation technology in Europe had become a serious enough problem that governments began to throw money at it. The most famous of these efforts was the United Kingdom's Longitude Act, passed by Parliament in 1714, offering the enormous sum of £20,000 to anyone who could devise a "practicable and useful" method of finding the longitude of a ship at sea to within half a degree.* The promise of financial reward stimulated no end of proposals, most of which seem crazy or impractical to modern readers,† but in the end two schemes were rewarded, one from each of the branches of timekeeping that we have discussed to this point.

The better known of the two was from the English clockmaker John Harrison, who developed a mechanical watch that could keep time at sea. The other large award handed out by the Board of Longitude was a posthumous £3,000 to German mathematician Tobias Mayer for a set of tables predicting the position of the moon with sufficient accuracy for navigational purposes. Mayer is far less celebrated than Harrison, but his method was in many ways the more immediately successful of the two, as his tables formed the basis of the *Nautical Almanac* that was distributed to ships by the British government and remained an essential navigational resource for a century and a half. Before Mayer could do his work, though, he needed another revolution in natural philosophy, one that changed the course of physics and mathematics forever.

THE SYSTEM OF THE WORLD: NEWTON AND HIS SUCCESSORS

According to popular legend, a young Isaac Newton was inspired to create his theory of gravity when he saw an apple fall from a tree on his family farm in Lincolnshire, in the East Midlands of England. This supposedly occurred in 1666 while he was sheltering from an outbreak of plague that closed the universities, which led to a lot of quickly played-out jokes about (re)inventing physics when the COVID-19 pandemic shut down schools and businesses in early 2020.‡

---

* Currency conversions at 300 years' remove are always dubious, but £20,000 would be equivalent to several million dollars in modern terms.
† And, to be fair, to many in Britain in the 1700s as well. Some of the more outlandish methods (e.g., using alchemical means to make a dog on a distant ship bark at noon London time every day) that are occasionally cited as real are, in fact, satirical presentations of longitude schemes as the domain of cranks and con artists.
‡ The first draft of this chapter was, in fact, written at home during this lockdown.

TIME AND NAVIGATION

The core problem of navigation on the earth is, in many ways, the same as the problem of locating stars and planets in the sky: you need to determine two angular positions, latitude and longitude. And just as with celestial coordinates, one of these is easy, while the other is very difficult.

The simpler coordinate is the latitude: as we discussed back in Chapter 1, latitude is easily determined from the height of the sun at noon or the height of the North Star at night. Given clear skies and a little practice, it's a straightforward process to find one's latitude, just as it's relatively easy to find the declination of an astronomical object.

Longitude, on the other hand, is difficult to find, for the same reason that right ascension is difficult to measure: there are no unambiguous reference points in the east-west direction. The constant rotation of the earth means that anything you sight in the sky will move from east to west at a steady rate, without hitting an obvious limit or landmark. Just as with right ascension, this rotation turns the measurement of longitude into a measurement of time. If you measure the time when some celestial object is directly overhead, and you know the time when it *was* overhead at another location to your east, you can determine the difference in longitude between those two points. The earth rotates through a full 360 degrees in 24 hours, so each hour difference in time is 15 degrees difference in longitude.*

A decade ago, when I first taught my class on the history of timekeeping, I arranged a set of simultaneous sundial measurements with Rhett Allain, a physics professor friend in Louisiana.† We each set up a simple sundial (Rhett's was a nail in a board, mine a stack of Lego bricks borrowed from my four-year-old), and recorded the shadow with a webcam. Reviewing the images afterward to identify the shortest shadows, we calculated our latitudes to be 43.4 and 31.3

---

* While this allows a good measurement of the longitude difference, it leaves open the question of what to use as a standard reference point when making maps. This is ultimately a political question—every seafaring nation tended to make its own maps referencing all longitudes to its own capital—and was ultimately resolved by political means, as we will soon see. The modern convention puts the zero of longitude in Greenwich, UK.

† He's also a prolific physics blogger, and the author of *Angry Birds Furious Forces* (National Geographic, 2013) and *Geek Physics* (Wiley, 2015).

degrees, in reasonably good agreement with the known values (from Google Maps) of 42.8 and 30.5. By comparing the time of solar noon in our two locations, we found our longitude difference to be 17.2 degrees, which is close enough to the correct value of 16.6 that it would've been worth say £15,000 from the Board of Longitude in the 1700s. The hardest part of the process by far was finding a day in January when it was sunny enough in both of our locations to obtain clear shadows.

This was easy for us because we knew the time of each image accurately, and we had modern communications technology that allowed us to share our data instantaneously across a distance of 1,400 miles. Neither of those would've been simple when the Longitude Act was passed, of course, which is why it was worth an enormous reward.

There were two basic approaches available in the 1700s to determine the time difference between two places on Earth. One was simply to set a high-quality clock to the correct local time at a reference location, then carry it to the destination whose longitude you wanted to determine. This was a challenge for mechanical clocks even on land in an era when carts pulled by animals were the fastest means of long-distance travel.

The other approach is astronomical and relies on observing a particular celestial event whose time can be predicted well in advance. The eclipses of the moons of Jupiter are perfect for this: as we saw in the previous chapter, their time can be predicted to within a few minutes, and they recur frequently enough to be convenient (unlike more spectacular but rarer events such as solar and lunar eclipses). This works very well on land—the surveying that irked Louis XIV was based in part on times determined from Jupiter's moons.

Jupiter is a small target, though, and observing the moons well enough to time the eclipses requires some skill even on land. On the deck of a ship at sea, it's basically impossible even for a trained astronomer. While Galileo and others attempted to construct head-mounted telescopes for use on ships, they never worked well.*

To be widely useful at sea, an astronomical method of determining the time would need a target that is easier to find and measure. The moon would seem to be ideal for this: as we discussed in Chapter 2, its cycle of phases makes it visible

---

* They do have an interesting sort of proto-steampunk aesthetic, though.

nearly every night for at least some time, and it's half a degree across when full, large enough to be easy to spot even on a wave-rocked ship. It also moves against the background stars rapidly enough—about half a degree per hour—to allow accurate timing during the course of a single night. And in fact, using the moon as a clock to determine longitude was first suggested by Johann Werner in 1514, a full two centuries before the Longitude Act.

Unfortunately for timekeeping, the orbit of the moon is exceptionally complicated, one of the most difficult to predict of any naked-eye object. As a result, developing a comprehensive theory of the lunar orbit with enough precision to be useful for navigation presented a formidable challenge to even the greatest scientific and mathematical minds of the day.

The Moon's Eccentric Orbit

It might well seem like the moon would be a simpler situation to deal with than any of the planets, as there is at least no ambiguity about what it orbits: in both the Ptolemaic and Copernican/Tychonic systems, the moon orbits the earth. And in a sense, it is simpler, in that it never shows the retrograde motion seen in the movement of the other planets. The moon always moves "forward" relative to the background stars, by a bit more than 13 degrees of arc per day on average,* never standing still or reversing direction. But while the general direction of the motion is relatively predictable, the *speed* varies significantly and in a complicated way. The *average* night-to-night shift is a bit more than 13 degrees, but for any given pair of nights, it can be higher or lower than that, depending on where it is in its orbit. The declination also moves back and forth in a band of several degrees above and below the ecliptic, again in a more complicated way than the simple cycle of monthly phases.

Astronomers have been tracking the behavior of the moon for thousands of years, though, and this long history of observations allowed them to identify several repeating cycles of different duration by as early as 500 BCE. In addition to the synodic month, the usual cycle of phases taking 29.530589 days,† they were aware of an "anomalistic month," the time between successive periods

---

* That's 52 minutes in right ascension if you use time units, but since the moon moves in both right ascension and declination, we'll stick with degrees, which can be used for both.

† All of the lots-of-decimal-places values given here are the modern values.

when the moon is moving fastest (27.554551 days), and a "draconic month," the time between crossings from south of the ecliptic to north of the ecliptic (27.212221 days). The colorful name for this last "month" comes from the fact that the "nodes," the points where the orbit of the moon crosses the region where the sun is seen in the sky, are the only positions in which an eclipse can occur,* and the oldest metaphor for an eclipse is a dragon swallowing the sun or moon. If this node crossing happens at the time of a full moon, the moon will pass through the earth's shadow, leading to a lunar eclipse. If the node crossing happens at the time of a new moon, the moon will be directly between the earth and the sun, and a solar eclipse can occur.†

These different lunar cycles are not simple multiples of one another, but over a long enough timescale, they do repeat, in much the same manner as the 19-year Metonic cycle that forms the basis of the Hebrew calendar we discussed in Chapter 2. At the end of 19 tropical years, the moon will have completed 235 synodic months, returning to the same phase on the equinox that it was at the start. In a similar vein, Babylonian astronomers noted that 223 synodic months is equal to 242 draconic months, leading to a pattern of eclipses separated by just over 18 tropical years.‡ This "Saros cycle" allowed the Babylonians and ancient Greeks to predict the times when eclipses were most likely to occur.

The cause of all this complexity comes from the fact that the moon is not describing a simple Keplerian ellipse under the influence of only the earth's gravity. In addition to the gravitational pull of the earth, the moon is also being pulled by a gravitational force from the sun, and this force varies with both the orbit of the moon around the earth (getting larger when the moon is between the earth and the sun, and smaller when it's on the far side) and the orbit of the earth around the sun (getting larger when the earth is closer to the sun, and smaller when it moves farther away). As a result of this perturbation, the

---

* Continuing the dragon theme, the ascending node where the moon crosses the ecliptic from south to north was traditionally called *caput draconis* (head of the dragon) and the descending node *cauda draconis* (tail of the dragon).

† Thanks to the size disparity between the earth and the moon, lunar eclipses last considerably longer than solar eclipses, and thus are more readily visible. The moon's shadow on the earth is relatively small, so solar eclipses are brief and only visible from a limited number of places.

‡ To be precise, 18 years, 11 days, and 8 hours.

moon's orbit around the earth does not close perfectly back on itself in a single ellipse, but instead traces out a complicated rosette pattern.

The gravitational interaction in the sun-earth-moon system is an example of a "three-body problem" in physics, and it's since been shown mathematically that an exact solution to predict the motion of all three bodies in such a situation is impossible.* Happily for astronomers, though, the changes in the force on the moon due to the sun are relatively small, and thus can be treated as a perturbation to a simpler solution. In much the same way that we can approximate a pendulum as a harmonic oscillator with a single oscillation period plus some small correction to the period that depends on the amplitude (the "circular error" discussed in Chapter 6), we can look at the motion of the moon as a simple orbit plus some additional corrections. More corrections are needed for the moon's orbit than a simple pendulum, but put all together they allow a relatively simple conceptual picture. The largest of these "lunar inequalities" had been identified by the 1600s, using a circular orbit, and acquired names like "Evection," "Parallactic Inequality," and the surprisingly generic "Variation."

Making a model good enough to enable accurate predictions months or years in advance, as needed for navigation in the "Age of Sail," requires a better description, the key elements of which were provided by Kepler's laws. To a very good approximation, we can treat the orbit of the moon as a Keplerian ellipse with the earth at one "focus," which is described by a fairly small number of parameters. The long axis of the ellipse—the "line of apsides" which contains the moon's closest approach ("perigee") and greatest distance ("apogee")—points along some direction in space that can be defined by a right ascension and declination. The moon's orbit lies in a plane that's tilted at some angle (about 5.1 degrees) relative to the orbit of the earth around the sun, and aligned so the node where the moon crosses from south to north points toward a particular location. The earth's orbit has its own line of apsides, and both elliptical orbits have an eccentricity, which tells you the difference between the shortest and longest distance from the focus of the orbit.

These nine parameters (three pairs of angular coordinates, one tilt angle, and two eccentricities) define the orbit at any given time, and most of the

---

* And that's even before you account for the gravitational interaction with all the *other* planets in the solar system, which make the situation even worse.

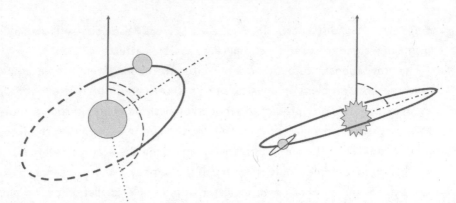

The orbital parameters for the moon. Left: Close-up on the earth-moon system, showing
the line of apsides (*dot-dash*) and the location of the node (*dotted line*) where the moon's
orbit moves from below (*dashed*) to above (*solid*) the path of the sun. Both lines are
described by an angle away from the origin of right ascension and declination (*arrow*).
Right: The earth-sun system showing the line of apsides for the earth's orbit (*dot-dash*)
and the inclination of the moon's orbit relative to the earth's orbit about the sun. (The
eccentricities are greatly exaggerated, and neither figure is remotely to scale.)

complexity of the moon's motion can be captured by allowing these parameters
to vary in time. The long axis of the moon's orbit rotates slowly, a phenomenon
called "apsidal precession," taking 8.85 years to complete a full loop. The orbital
plane also rotates, carrying the orbit with it, so that the position of the nodes
makes a complete cycle every 18.61 years ("nodal precession"). The eccentricity
of the moon's orbit varies depending on where the earth is in *its* orbit, becoming
more stretched when the earth is closer to the sun, and more round when the
earth is farther away. And so on.

   The challenge of developing a useful lunar theory is to predict how the
parameters that describe the moon's orbit are changing at any particular
moment. This is partly a matter of making careful empirical observations—
several of the important corrections to the simple theory of the moon's orbit
were first identified thanks to decades of observations recorded by Tycho
Brahe. Extending this into the future with sufficient precision to be useful
for timekeeping and navigation, though, requires some deeper understanding
of what's causing the orbital perturbations in the first place, which was only
possible thanks to Newton's revolutionary work on the physics of gravity.

There are a great many problems with the falling-apple story, not least that it doesn't appear until decades later when a much older Newton told it to friends and family.* The most fundamental problem, though, is that the popular version misstates the nature of Newton's insight. His key realization was not that gravity pulls things to the earth—it's not like everyone was floating around weightless prior to 1666, after all. Gravity was well known and had been extensively studied by Galileo, Riccioli, Flemish polymath Simon Stevin, Dutch philosopher Isaac Beekman, and others. Newton's actual flash of inspiration was that the gravitational force pulling an apple to the earth could be the *same force* that holds the moon in its orbit.

Prior to Newton, the question of what drove the motion of celestial bodies was an enormous problem for philosophers. The oldest models of the universe had the stars and planets supported in a physical way, borne by vast "crystal spheres," but observations by Brahe and others showing that comets crossed through the orbits of the planets ruled out any such arrangement. Some Aristotelian-derived models classified circular motion as a "natural" state that could go on forever without interruption, but this is difficult to square with Kepler's spectacularly successful model of elliptical orbits. Kepler himself became convinced that the planets must be held in orbit by some force emanating from the sun, but he remained at a loss as to what that force might be.

This is the very question that, whatever the time and place of his initial insight, finally led Newton to publish his ideas about physics, thanks to a wager among three other early scientists. In early 1684, Edmond Halley, Christopher Wren, and Robert Hooke had a friendly discussion about the properties of the force holding the planets in their orbits. All three suspected that the force experienced by a planet in orbit would need to decrease like the square of the distance (i.e., a planet twice as far from the sun would feel one-fourth the force).† It was not immediately clear, though, that such a force would necessarily lead to Keplerian elliptical obits. When Halley raised the question, Hooke claimed to

---

* This came after years of bitter arguments with the German mathematician Gottfried Leibniz and others about who was the first to invent key elements of calculus and physics; placing his inspiration in 1666 conveniently supported Newton's campaign to establish priority.
† They were led to this idea by Christiaan Huygens's work on how centrifugal forces vary with speed and distance, and the relationship between orbital speed and distance found in Kepler's laws.

have a proof of this that he was keeping secret, which was fairly typical behavior for him. Wren offered a prize of 40 shillings worth of books to whichever of them could come up with a proof in the next several months.

That August, Halley visited Newton, who was then at the University of Cambridge and known to be both brilliant and reclusive, and asked him what shape of orbit would be expected for a planet attracted to the sun by such an inverse-square force. Newton immediately replied that the orbit would be an ellipse, saying that he had already calculated it. Halley was amazed, and asked to hear more, but Newton was unable to find the proof that day. He promised to redo it and send it to Halley.

Newton's perfectionist tendencies turned this into a process taking months, but when his discussion of orbital motion arrived in November, Halley was astonished at how comprehensive it was, hinting at a fully developed new science of mechanics. Halley urged Newton to publish his work, using his position as secretary of the Royal Society to secure a promise of support. Thus began a process that culminated in the publication of *Philosophiæ Naturalis Principia Mathematica* (Latin for "Mathematical Principles of Natural Philosophy"), now commonly known as "Newton's *Principia*."

The writing and publication of the *Principia* was not a smooth process, and it stretched Halley's resources to the limits. Newton was a difficult person to deal with at the best of times, needing constant gentle encouragement, and publication was almost derailed completely when Robert Hooke incensed Newton by claiming to have found the inverse-square form of gravity before Newton.* Thanks largely to Halley, though, the *Principia* finally saw print in 1687. The final book is a monumental work, published in three volumes, laying out Newton's laws of motion and theory of universal gravitation and applying them to a wide range of systems.

In the first book of the *Principia*, Newton spelled out his laws of motion: that objects move in straight lines at constant speed unless acted upon by an external force; that the acceleration of an object is equal to the force acting on

---

* Halley managed to calm Newton down enough to finish the book, but then wound up having to cover the printing costs out of his own pocket. The Royal Society's budget was in bad shape after the unsuccessful publication of a massive illustrated book called *A History of Fishes* the year before. Halley was eventually compensated for his work bringing Newton's masterpiece to print, but his pay came in the form of unsold copies of *A History of Fishes*.

it divided by the mass; and that any object that exerts a force on another object experiences a reaction force of the same size in the opposite direction. He also introduced a mathematical expression for the force of gravity between any two objects with mass, which is proportional to the product of the masses and has the inverse-square distance dependence Halley had suspected. Starting from these principles, Newton showed how they lead to Kepler's laws for the orbits of planets.

The second book dealt largely with the motion of objects in fluids such as air or water, including the effects of air resistance on the motion of a pendulum, though this is not without an astronomical connection. Newton's study of fluid motion was undertaken in part to refute a theory, put forth by René Descartes, that the planets were carried in orbit around the sun by swirling vortices in a fluid that filled the space between planets. This vortex theory was influential for a time, particularly among those who were bothered by the way Newtonian gravity acted over vast distances with no obvious medium to convey the force. In the end, though, the precise quantitative predictions possible with Newtonian gravity, and the inconsistency of the vortex model with the behavior of real fluids, carried the day.

In the third book, *De mundi systemate* ("On the System of the World"), Newton turned to detailed applications of the principles set forth in the first two, and compared them directly to observational data. A considerable amount of this book was devoted to exploring the orbit of the moon: Newton laid out the basic idea of the three-body interactions between the sun, the earth, and the moon, and showed how the changing magnitude and direction of the sun's gravity on the moon can give rise to some of the observed perturbations of the lunar orbit. The quantitative calculation, though, does not work out as well as he might have liked: while Newton had the basic conceptual picture of how the force from the sun causes the apsidal precession of the moon's orbit, his calculation of the rate of precession was half as big as that observed.

Given the complexity of the problem, it's amazing that Newton's theory got even this close, but the factor-of-two discrepancy was a distinct problem, and it led to speculation that the inverse-square mathematical form of gravity was not, in fact, universal, and might need to be modified in some way. Some of Newton's rivals, particularly on the European continent, saw this as a potential weak point from which to start a more comprehensive attack on his physics.

In the end, though, the problem is just that these calculations are ferociously difficult, even for the greatest mathematicians of the time. After many false starts, the solution was found, first by the French mathematician and astronomer Alexis Claude Clairaut, closely followed by the great Swiss mathematician Leonhard Euler. While Clairaut and Euler had each started out believing that Newton's inverse-square law would need to be modified to explain the orbit of the moon, by the early 1750s (a quarter century after Newton's death) both had succeeded in showing that all the observed perturbations of the moon's orbit could be satisfactorily explained using Newton's theory.

Between Newton's new physics and the mathematical theories of Clairaut and Euler, it was at least in principle possible to predict the orbit of the moon well enough for it to serve as a clock. All that remained was the hard and decidedly unglamorous work of carrying through the many calculations needed to convert the equations for the orbit into a set of tables predicting the position of the moon.

TOBIAS MAYER AND THE MOON

Johann Tobias Mayer was born in 1723 near Stuttgart. His father was trained as a wainwright but demonstrated expertise with a broader range of mechanical things, which won him a commission to build and manage water systems in Esslingen when Tobias was still an infant. As a child, Mayer imitated his father's disassembling and sketching of mechanical devices, developing mechanical and artistic abilities that would serve him well in later life.

Mayer's father died in 1731, leaving Tobias and his siblings in dire financial straits, but through the mentorship of the mayor of Esslingen and a mathematically inclined local shoemaker,* Mayer was able to continue his education. He developed a strong interest in becoming an artillery officer, and to that end spent a good deal of time learning to map and draw fortifications and do the geometrical calculations needed to determine the range and charge needed to hit a chosen target. While he never did manage to secure a military commission,

---

* The shoemaker would buy treatises on mathematics, which Mayer would read during the day while his friend was making shoes, and they would discuss them together after dinner.

this education led to work as a cartographer, and he eventually settled in Nurem-
berg in 1746. He joined the Homann Cartographic Bureau under its director
Johann Franz* and began work on a new atlas. He was distressed to learn, however, just how bad the state of cartography was: only 22 locations in Germany
had their latitude firmly established by astronomical measurements, and only
139 places on the entire Earth had both latitude and longitude reliably recorded
in the sources available at the Homann Bureau. This shortage of information
triggered an interest in astronomy, as Mayer set out to improve the techniques
for determining earthbound coordinates by looking to the sky.

In pursuit of a method that would be widely applicable, Mayer began
studying the moon carefully and mapped the lunar surface more precisely than
anyone else had done to that time. He determined that the moon had no atmosphere (based on the lack of visible distortion as the moon passed in front of a
background star) and made the best measurements to that point of the "libration" of the moon. It's well known that the moon's rotational period matches
its orbital period, so that the same hemisphere of the moon is always facing
the earth, but because the moon's orbit is elliptical, this isn't quite perfect. The
visible surface wobbles back and forth by a few degrees of lunar longitude or
latitude over the course of an orbit; as a result, the portion of the lunar surface
that's *ever* visible from the earth is about 59 percent of the total. Between 1748
and 1749, Mayer made repeated careful measurements of the angular distance
between the edge of the visible moon and prominent features on its surface, and
determined the amount of this libration.

Mayer's ultimate goal lay in the area of mapmaking, specifically attempting
to develop a method of predicting longitude based on the motion of the moon.
In 1751, he moved to Göttingen to take a professorship and began a correspon-
dence with Leonhard Euler, who was then putting the finishing touches on his
model of the moon's orbit. Armed with Euler's formulae and his own observations, Mayer began the tedious task of calculating the future positions of the
moon, and by 1753 he had a set of lunar tables accurate to within 1.5 minutes
of arc. He continued observing and calculating, and within two years reduced
this uncertainty to under a minute.

---

* In addition to being his new boss, Franz was a future relative: he introduced Mayer to his
sister-in-law, Maria Gnug, whom Mayer married in 1751.

A one-minute-of-arc error in the position of the moon is sufficient to determine time well enough to calculate longitude within half a degree, meeting the qualification for the £20,000 prize offered in Britain. The urging of friends and colleagues (notably Euler) overcame Mayer's justified skepticism that the British Board of Longitude would look favorably on a submission from a foreigner, and he sent a copy of his tables to London in 1755. This lunar method was tested by the astronomer Nevil Maskelyne on a voyage to Barbados in 1761, measuring "lunar distances" between the moon and other celestial objects with well-known positions. These distances were then compared to the predicted positions based on Mayer's tables and used to work out the time.

Maskelyne found that the lunar method worked very well, allowing him to determine the longitude to within the targets set by the Longitude Act; unfortunately, the calculations involved were very complicated. He realized, though, that much of the work could be done in advance, adapting Mayer's tables from a form best suited for scientists who had the skills and time to something simpler and faster for shipboard use. On his return to England, he dedicated himself to this operation, assembling a team of calculators to convert Mayer's work to a more sailor-friendly form.

Following Maskelyne's report and the receipt of an improved set of tables sent by Mayer the following year, the Board of Longitude eventually awarded £3,000 for his accomplishments in 1765. Sadly, Mayer himself had died in 1762, at age 39, so the money went to his widow and children.* Mayer died young enough to make one wonder what he might've accomplished given more time, but he did achieve a form of immortality via his life's work. Mayer's lunar tables formed the basis of the *Nautical Almanac* assembled by Maskelyne and published at the Royal Observatory. These tables made reliable determinations of longitude widely available for the first time and were rightly celebrated as an enormous boon to sailors the world over. The *Nautical Almanac* has been updated and distributed continuously since 1767 (with the US Naval Observatory putting out its own analogous tables since the mid-1800s), and even in the modern era finding the time by astronomical methods is taught at the US Naval and Merchant Marine academies. The work of refining and updating

---

* An additional, unsolicited award of £300 went to Euler for developing the mathematics behind Mayer's tables, proving once again that it's good to be famous.

calculations of the positions of celestial objects continues as well, and for a brief period in the twentieth century even served as the basis for the definition of the second (as we'll see in Chapter 13). Before we get there, though, we need to discuss the other great technological innovation supported by the Board of Longitude, which led in turn to a change in our everyday experience of time.

Chapter 10

WATCH THIS

One of the most universal characteristics of the passage of time is that it's *not* universal: everybody experiences it a little bit differently. As we'll soon see, there's a physics sense in which that's quantifiably true, but most of the differences between our experiences of time are subjective. A full day of work can seem to fly by or drag on endlessly, depending on how you feel about the tasks at hand, and what seems to a parent like a perfectly reasonable wait for an amusement-park ride will feel like HOURS to their young children.

This difference in subjective experience is exacerbated by the fact that, particularly in the adult world, everybody keeps their own schedule. The person at the supermarket checkout fumbling to find the exact change for their purchases might see this as a good use of time because they have no pressing responsibilities at the moment, while the harried person behind them is steaming because they're running late for a meeting. Everybody's doing their own thing, at their own pace, and when schedules bump up against each other it can create tension.

These differences in subjective experiences of time create a demand for *objective* time that is nonetheless individual, so we can each carry around a reference to a shared and synchronized official time. This allows us to verify that, no, we have not been in this line for a literal hour, and yes, there's still time to get to the office. These days, this mostly takes the form of various calendar apps on smartphones and computers, and before that it was wristwatches and paper planners. The impulse goes back basically as far as there were standardized hours of the day: archaeologists have found numerous examples of portable pocket sundials dating to Roman times.

The demand for portable time increased dramatically after the invention of mechanical clocks, which have the distinct advantage of (at least potentially) working well in any weather, and also indoors where the sun doesn't shine. As a result, the development of small chamber clocks and pocket watches closely parallels that of church tower clocks and other monumental timekeepers.

Up until the 1700s, though, these mostly weren't very good, particularly the watches. Elaborate clocks and watches were valued less as accurate time-keepers than as symbols of wealth and status. Announcing the fact that you could afford to carry around an expensive watch and use it to order your day mattered more than whether the time it showed was actually correct.

The longitude problem adds a new element to the demand for portable time. For ships at sea, it matters very much that their clocks show the *correct* time, as a reference for determining longitude. The method of lunar distances works well to accurately determine longitude at sea, but requires a bit of calculation, even using the simplified tables of the *Nautical Almanac*; a reliable clock dramatically simplifies the process. A sailing ship at sea is an incredibly unforgiving environment for a mechanical device, though, so the problem of making an accurate marine chronometer posed a formidable challenge. Even with the incentive of the Longitude Prize, it took decades of work to construct a portable timekeeper that could stand up to the rigors of sea travel.

MARINE CHRONOMETERS

Calculating lunar distances was the first widely used method for correctly tracking longitude at sea, but it's not the only solution developed in the mid-1700s. The other approach to finding the longitude, more celebrated in modern times, now seems far more obvious and straightforward: simply make a clock that keeps accurate time and carry it with you as you travel. If you have a clock that shows the time at the place you left, it's a simple matter to determine the difference in longitude between there and your current location.

THE EARLY DAYS OF PORTABLE TIME

Almost as soon as the first mechanical clocks were invented and installed in church towers in Europe, there was a demand for smaller clocks that could be installed and operated indoors. The medieval French poem *Le Roman de la Rose* contains references to chiming chamber clocks in sections dating from 1275 CE, around the time a (presumably mechanical) clock was installed above the rood screen in Dunstable Priory. Records of the estate of French king Charles V in 1380 show that it included a small silver clock originally made for his ancestor Philip the Fair, who died in 1314.

The making of chamber clocks posed a number of challenges for the technology of the day, beginning with the availability of materials. Medieval tower clocks were mostly made from cast iron, which is sturdy enough to hold up to the repeated impacts involved in the operation of a verge-and-foliot clock, but the ironworking technology of the day was not up to fine work at small scales. Gears small enough for an indoor clock needed to be cut from brass or other metals less durable than iron, so producing a brass clock with a reasonable lifetime required reengineering the gear train to reduce stress. The smaller works also magnify errors introduced by imperfections in the gears—an error of a millimeter is insignificant in a tower clock gear half a meter across, but it can significantly reduce the accuracy of a chamber clock with gears just a few centimeters in diameter.

Chamber clocks also posed a problem for the drive technology. A clock driven by falling weights can run a long time when placed at the top of a tower, or mounted high on a wall, but this design is not well suited to a tabletop device, let alone a watch. Making a compact clock requires a different drive technology, one that can fit in a smaller space but still store enough energy to run for a day or more.

By the early 1400s, a new drive technology had emerged: a coiled metal spring wound up so as to store energy and slowly release it through the drive train. These brought yet a new problem, though: as we learned when we talked about Hooke's law in Chapter 6, the force delivered by a spring is not constant but increases as more stress is placed on it. As the mainspring unwinds, it supplies less force to the drive, so a spring-driven verge-and-foliot clock ticks more slowly as it winds down. This problem was addressed by the invention of the "fusee," a conical wheel mounted on the drive shaft. In this version, the mainspring no longer pushes directly on the drive shaft but instead is connected by a fine chain wound around the fusee, using the physics of torque to compensate for the decreasing force as the spring unwinds. When the spring is fully wound and pulling with its strongest force, the chain is pulling on the narrowest part of the cone, thus producing a relatively small torque to rotate the drive shaft. As the clock runs, the spring unwinds and so does the chain, so the force is applied at wider parts of the fusee at later times. The greater distance from the center compensates for the weaker force from the spring, just as using a wrench with a longer handle produces a larger torque when trying to loosen a particularly difficult bolt. The changing radius of the fusee allows the decreasing force from the mainspring to produce the same torque on the drive shaft, keeping the clock ticking at the same rate.

The fusee was invented in the early 1400s—the earliest surviving example is in a clock built for Philip the Good, the Duke of Burgundy, in 1430—enabling the production of more accurate clocks at much smaller scales. In the hands of skilled artisans, an impressive degree of miniaturization was achieved. Francis I of France is on record as buying two daggers with tiny clocks built into their hilts in 1518, and Elizabeth I of England had a clock built into a finger ring, including an alarm feature using a small prong to scratch her.

Spring-driven clocks with a fusee could also be combined with a remontoire, a system in which the drive force for the clock was provided by a small spring, which was periodically rewound using the force of falling weights or a larger mainspring. These innovations powered high-precision clocks before the invention of the pendulum clock: the best verge-and-foliot clocks ever built, by the Swiss clockmaker Jost Bürgi in the early 1600s, were good to around a minute per day.

The fusee and remontoire are readily adaptable for use in pendulum clocks—if anything, they're even better suited to a pendulum clock than one with a foliot, because the force required is much lower. This allows for the construction of compact and high-precision pendulum-based chamber clocks, which could readily be transported from one place to another and set back up to provide an accurate time reference. Making a clock that would keep time *during the trip*, however, was a much more difficult matter, and one that has a direct bearing on the longitude problem we have been discussing.

This approach to the problem of longitude was proposed not long after the method of lunar distances by the Dutch mathematician and cartographer Gemma Frisius in 1530. While it's simple in concept, actually making a sufficiently accurate clock is a formidable challenge, well beyond the clock technology of Frisius's day. Even a hundred years later, the French astronomer Jean-Baptiste Morin dismissed the clock method out of hand, saying, "I do not know if the Devil will succeed in making a longitude timekeeper, but it is folly for man to try."*

Morin's sweeping declaration did not stop people from trying, of course. Almost as soon as the pendulum clock appeared on the scene, scientists and clockmakers began trying to make a clock for finding longitude. Christiaan Huygens constructed a pendulum-based marine chronometer in the early 1660s, and some early tests appeared promising. These were not borne out by tests on longer voyages, though. In 1670, Jean Richer was in charge of a seagoing test of Huygens's pendulum clocks, but the trial was abandoned after the clocks stopped running. Huygens initially blamed this on Richer's carelessness, but further tests in 1686–1687 and 1690–1692 fared no better, and Huygens eventually abandoned the project.

As it turns out, shipboard is a singularly awful environment for a pendulum-based mechanical clock. The regular swinging of a pendulum depends on the downward pull of gravity, but on the deck of a wave-rocked ship, the exact orientation of the clock changes constantly, introducing errors in the rate of ticking. Effectively, the rocking introduces extra "circular error," because each time the pendulum reaches its turning point, it's at a different angle from the true vertical. Even if this could be countered, a seagoing pendulum clock must also contend with the problem discovered by Richer in Cayenne: the strength of gravity is different at different latitudes, changing the period of a pendulum by enough to introduce errors of a few minutes per day. This manifested as an apparent eastward shift in longitudes measured on the 1687 voyage of Huygens's marine clock, which ticked slower as it approached the equator.

---

* Quoted by Derek Howse in "The Lunar-Distance Method of Measuring Longitude," in the proceedings of the 1993 Longitude Symposium. W. J. H. Andrews, ed. *The Quest for Longitude.* Harvard University, 1996.

The pendulum-specific problems can be addressed by removing gravity from the equation, by replacing the pendulum with a "balance spring" consisting of a metal ring attached to a spiral spring. The ring twists back and forth, and this motion provides the regulation for the drive. Several people tried to claim credit for the idea, among them Huygens, the Parisian clockmaker Isaac Thuret, the French physicist Jean de Hautefeuille and, inevitably, Robert Hooke; while all of them had at least pieces of the idea, there's little dispute that Huygens directed the building of the first balance-spring watch.*

The balance spring may seem, at first glance, to reintroduce the problems of the foliot, with the rate at which the metal wheel twists back and forth depending on how hard it's pushed by the drive force. The spring takes care of this, though, because it supplies essentially all of the force needed to stop and reverse the motion: just as in a pendulum clock, the mainspring drive only needs to provide a small force to sustain the motion against the energy loss due to friction. Balance spring regulators can be remarkably stable and are still used in modern mechanical watches.

While the balance spring is a sound idea in principle, it faces severe problems in the real world, particularly in the context of navigation at sea. A sea voyage in the 1700s was an incredibly harsh environment for any mechanical device, with temperature and humidity varying over huge ranges. The temperature variation would change the rate of ticking as the ring for the balance expanded and contracted. More than that, though, corrosion of metal parts due to the ever-present salt spray would degrade the performance of a clock (or anything else mechanical) by increasing the friction between moving parts, making it harder to turn gears and generally gumming up the works. This could to some degree be mitigated by lubrication, but the lubricants available in the 1700s were prone to turning sticky outside of a narrow range of temperature, so in many ways they only made the problems worse.

By the early 1700s, the core technology needed to make accurate chamber clocks and even watches had been developed and deployed in observatories all around Europe. While these clocks performed well on land, the harsh marine

---

* The balance spring is also ideal for use in pocket watches, which were never able to employ pendulums as regulators, since the very nature of a watch is to be carried around in a variety of different orientations. Early watches always used some version of a foliot as a regulator, which did not help with their accuracy issues.

environment continued to defeat the best clockmakers, and Morin's dismissal of the idea of a longitude timekeeper continued to seem prescient.

CLOCKS TAKE TO THE SEAS

The problem of making an accurate marine chronometer was eventually solved by John Harrison, an English clockmaker from Lincolnshire. Harrison was trained as a carpenter but had always been fascinated by clocks and their operation,* and despite being self-taught became one of the unquestioned masters of the clockmaking craft. Harrison first gained notice for his longcase clocks, including a "self-lubricating" clock with wooden gears,† and the invention of the gridiron pendulum to counteract thermal expansion that would otherwise slow the clock as the temperature increased (as discussed in Chapter 6). Around 1730, he decided to build a marine chronometer to win the Longitude Prize, a project that was to consume the rest of his life.

Harrison's first marine chronometer, now dubbed "H1," used an elaborate "grasshopper" escapement of his own invention, and twin barbell-shaped balances connected by springs. These oscillated in opposite directions to counter the effect of rocking motion. A replica of H1 can be seen in operation at the Royal Museum in Greenwich, UK, and it's beautiful in the way that intricate mechanical devices often are.

Harrison took H1 on a voyage to Lisbon and back in 1736; after some breaking-in issues on the outward leg, it performed extremely well on the return trip.‡ This was encouraging enough that the Board of Longitude awarded him £500 to support further work; several years later, this was followed by another £500 payment. Nearly 20 years passed before the next seagoing trial of a Harrison clock, though, during which time he drastically revised the design.

---

* According to one story, while Harrison was recovering from smallpox as a child, he disassembled and reassembled a watch to figure out how it worked.

† The particular wood Harrison used, lignum vitae, is very dense, with a high content of natural oils that maintain a very smooth, low-friction surface.

‡ In addition to learning about his clock's performance, Harrison also learned that he was prone to seasickness, and after this trip never went abroad again.

Where H1 is an intricate, almost showy device, its successor H4*—the chronometer that finally met the standards laid out by the Longitude Act—looks like nothing more than a slightly larger than normal pocket watch. Inside its case, though, it hides a number of significant innovations.

By the time he made H4, Harrison had switched to using a balance spring regulator, which, as noted above, removes the dependence on the strength of gravity but does face significant problems in dealing with changes in temperature. Harrison found an ingenious solution to this problem: he included a compensation curb made out of a "bimetallic strip," a piece of brass riveted together with a strip of steel. Like the gridiron pendulum, this balances the different rates of thermal expansion of two different metals to keep the oscillation period constant. Both metals expand as the temperature increases, but the brass side expands more rapidly, causing the strip to bend toward the steel side. The bending of the curb causes it to press against the spring, effectively shortening its length and increasing the force to compensate for the slower oscillation due to the expanding wheel.†

In addition to the bimetallic curb, H4 is masterfully engineered and constructed.‡ The escapement of H4 is a variant of the verge, but redesigned in a subtle way that has less recoil. The pallets of this escapement are also embedded with chips of diamond on the contact surfaces. Diamond provides a smoother and more durable surface than could be made with metal at that time, so these chips reduce the friction of the works with no need for lubrication.

H4 went to sea in November of 1761, accompanied by Harrison's son William, for an 80-day voyage to Barbados—the same voyage where Mayer's lunar tables were tested. On arriving in the Caribbean, H4 showed a time within about five seconds of that expected from the known longitude difference. This corresponds to a position error of around one mile, much better than the

---

* The prototype clocks H2 and H3 ultimately didn't meet Harrison's exacting standards and never went to sea.
† Later watches use Harrison's bimetallic technology on the balance wheel itself, dividing the wheel into semicircular strips with a small gap between them. When the temperature increases, the strips bend to curve more tightly, keeping the size of the wheel (and thus its rate of oscillation) constant.
‡ And also beautifully decorated, in the manner of the day, unlike the strictly utilitarian appearance of modern engineering prototypes.

standard set by the Longitude Act. A repeat trial in 1765 was accurate to within about 10 miles, again, much better than formally required.

To the Harrisons's great annoyance, rather than being awarded the Longitude Prize, this was instead the start of another near decade of wrangling with the Board of Longitude and the Astronomer Royal, Nevil Maskelyne. The Harrisons felt that Maskelyne, as an astronomer, was unfairly biased against clocks in general, a feeling that was heightened when Maskelyne began producing and distributing the *Nautical Almanac* based on Mayer's lunar tables. Harrison was never officially awarded the £20,000 Longitude Prize, though he was paid £10,000 for the design of H4 and awarded another £8,750 by a special act of Parliament after a personal appeal to King George III.*

The many twists and turns of this process have been well chronicled, most famously in Dava Sobel's prize-winning book *Longitude*. This book very much takes Harrison's side and casts Maskelyne as the villain, but on careful examination,[†] the case is not so clear. The ultimate argument turns on the "practicable and useful" requirement of the Longitude Act: Harrison's chronometer solved the technical problem of tracking longitude but required heroic effort and enormous expense. In the late 1700s, a single quality marine chronometer cost something like £40, while a reliable quadrant and the tables of Maskelyne's *Nautical Almanac* could be purchased for a tenth of that. Additionally, while keeping time with a clock allows one to *track* longitude in a way that doesn't require astronomical measurements and mathematical calculations, it can't be used to *find* longitude unless the clock is running continuously. Once the clock stops ticking, it also stops being a useful navigational instrument until it can be reset to the correct time. While at sea, it can only be reset through something like the method of lunar distances.

---

* Including the various development grants he received over the years, Harrison ended up getting a bit more than the £20,000 all told but never received the official recognition of being declared the winner of the Longitude Prize.
† Sobel's book was originally published by Walker & Company in 1995 and reissued by Bloomsbury Publishing in 2005. A more comprehensive treatment of the topic can be found in the book *Ships, Clocks, and Stars: The Quest for Longitude* by Richard Dunn and Rebekah Higgitt (Harper Design, 2014), which was written to accompany a major exhibition at the National Maritime Museum in Greenwich, UK.

For that reason, even as the price of chronometers came down and their reliability improved, astronomical techniques remained essential for seagoing navigation for more than a century after Harrison built his clocks. In general practice, though, as quality mechanical clocks became more common and more reliable, the need for "taking lunars" to find longitude decreased, so by the end of the 1800s it was rarely used. With the introduction of radio in the early years of the twentieth century, allowing time signals to be broadcast over huge distances, celestial navigation almost completely disappeared from the everyday operation of ships at sea, and even shipboard clocks were greatly reduced in importance.

From Ship to Shore

One of the key points of contention in the wrangling between Harrison and the Board of Longitude was the latter's demand that Harrison show other clockmakers how he built H4. Harrison denounced this as an outrageous imposition on his prerogatives as an inventor and clockmaker, but from the standpoint of the "practicable and useful" requirement, it makes perfect sense. Harrison was not a young man—he was almost 70 years old by the time H4 went to sea—and a watch that only he could build could not address the scale of the problem of merchant shipping.

In 1765, Harrison finally agreed to disassemble and reassemble H4 in front of a team of experts selected by the board, one of whom was watchmaker Larcum Kendall, who was then assigned to make a copy of H4. Kendall's first watch, known to historians as "K1," was carried by Captain James Cook on his second South Seas voyage. "Kendall's watch has exceeded the expectations of its most zealous advocate,"* reported Cook. This endorsement helped firmly establish the idea of chronometers as a viable navigational method.

While K1 performed marvelously well for Cook on that voyage, it was still enormously expensive—it cost £450 to build,† a good fraction of the price of a ship, and took nearly four years. To be truly practical, the watch would need to be simplified to allow faster manufacture. Kendall himself made two additional watches based on modifications of Harrison's design (K2 and K3), though these were markedly less accurate than the original copy.

---

* J. C. Beaglehole, *The Life of Captain James Cook*. Stanford University Press, 1974.
† Kendall received a bonus of £50 when it was completed.

The real work of expanding chronometer production to the necessary scale fell to other clockmakers, who built on Harrison's principles but incorporated new types of escapements. The most significant of these variations came from John Arnold and Thomas Earnshaw (who, inevitably, spent several years involved in a patent dispute over the invention of a particular type of escapement). The historian David Landes estimates that by 1815 Arnold, Earnshaw, and their colleague Paul Philip Barraud had produced a couple thousand high-quality clocks using these techniques.

While a first-rate marine chronometer remained an expensive item, the same techniques were readily adapted to the more forgiving environment on land at much lower cost. The watchmaking operations of Arnold and Earnshaw turned out many more civilian watches than marine chronometers, and even these are remarkable and robust devices. As we'll see in the next chapter, one of Arnold's watches from 1794 remained in daily use up until World War II, transporting time from the Royal Observatory to various businesses around London. And, of course, there were scores of lesser clockmakers producing watches that, while not to the standard of Arnold and Earnshaw, were more than adequate for the purpose of coordinating daily activities. By the mid-1800s, reliable mechanical watches were moving portable time toward the mass consumer quantity that it is today.

GENIUSES AND CRAFTSMEN

Many of the stories we tell about the historical development of science have a tendency to focus on the work of "geniuses" like Isaac Newton, figures whose discoveries or new theories have a revolutionary impact. The subtitle of Dava Sobel's book aside, though,* both clockmaker John Harrison and astronomer Tobias Mayer are more properly regarded as master craftsmen: the primary cause of their success was painstaking, unglamorous work.

Mayer's lunar tables were founded on the work of mathematical geniuses like Newton and Euler, who developed the physical and mathematical formulae needed to predict the moon's complicated orbit. While Newton and Euler

---

* "The True Story of a Lone Genius Who Solved the Greatest Scientific Problem of His Time."

established the principles behind the method, it was Mayer who did the labor to make the tables a reality: enumerating and evaluating the various perturbations affecting the orbit to determine which were most important, and matching the formulae to data from years of careful observations. Euler and Clairaut showed that Newton's concepts of physics *could* predict the orbit, but Mayer turned that concept into reality.

Harrison, for his part, succeeded largely because of the messy science of materials. His clocks include technical innovations, most notably the use of bimetallic strips to compensate for temperature changes, as mentioned, but their key elements—balance springs, remontoires, fusees—were well known before Harrison. He raised their use to a high art by being incredibly exacting in his choice and use of materials: finding the perfect combination of metals and incorporating diamond pallets to reduce friction. The two decades that passed between H1 and H4 were spent piecing this together by trial and error.

The stories of Mayer and Harrison are a reminder that the most important milestones in history draw from both sides of the (somewhat arbitrary) divide between science and engineering. Their achievements would not have been possible without the theories developed by more celebrated scientists, but turning theory into practice is a long and arduous process. Mayer and Harrison have fared better than many—their names are at least recorded and remembered, unlike those of the anonymous innovators who made the first sandglass and the first verge-and-foliot clock.

The successful resolution of the longitude problem, thanks to the *Nautical Almanac* and seagoing clocks, was a great boon to the growth of globe-spanning empires, helping establish and maintain the reliable long-distance shipping networks that brought enormous wealth and political power to European capitals. The next key innovations in timekeeping also had to do with long-distance transportation, in this case the rapid movement of people and information in a way that allowed people to experience the change of time that comes with a change in longitude in a direct and immediate way. Resolving *that* problem would completely transform the way we conceive of time in a global sense, thanks in large part to contributions from two more underappreciated groups: bureaucrats and politicians.

Chapter 11

DOES ANYBODY REALLY KNOW WHAT TIME IT IS?

For the vast majority of human history, time was a local phenomenon. A day was defined by the rising and setting of the sun at a particular place, and even as time began to be formalized by public clocks, it still remained tied to the motion of the sun. Each town would set its own clocks to synchronize with the sun so local noon coincided with the sun reaching its highest point. At noon in Washington, DC, a clock in New York City would show 12:12 PM, while a clock in Charleston, South Carolina, showed just 11:48 AM.

This meant that knowing what time it was in a distant town was a difficult problem, requiring detailed information about the difference in longitude between the two endpoints. As a practical matter, though, the question was largely moot. The sun might rise and set later at a traveler's eventual destination, but given the available means of transportation, this had no appreciable effect. A person on a horse might reasonably expect to cover 30 miles in a day; this corresponds to a solar time difference of maybe three minutes at North American latitudes. That small a shift would hardly make an impression, particularly after a long day in the saddle.

By the mid-1800s, though, this had changed significantly: a passenger on a train might reasonably expect to travel 30 miles in an *hour*. A traveler leaving

New York City by train in the years just before the Civil War could reach the middle of Ohio in the course of a single day; the corresponding time change would be more than half an hour, a readily noticeable shift in the time of sunset (particularly given the increasing availability of reliable watches during the same period). Differences in local time began to take on more practical importance for people on the move.

As much as the speed of people had changed by 1860, the speed of information had changed far more. The spread of the railroads was accompanied by the growth of the telegraph, providing the first truly worldwide telecommunications network. The first telegraph line connecting the coasts of North America was completed in 1861, and reliable transatlantic telegraph service was established by 1865. Suddenly, messages could be sent from San Francisco to London and back in a single day, a leap in communications that required a more careful consideration of time on a global scale.

This rapid progress in communications and transportation eventually led to our modern system of standardized time zones, where all cities in particular bands of longitude set their clocks to the same time. This decouples time from the motion of the sun—if solar noon coincides with 12 o'clock in Washington, DC, the sun is already 12 minutes past its peak in New York, and still 12 minutes short in Charleston—but it simplifies the process of moving people and information over long distances. The move to separate clock time from solar time is an important reminder that for everyday purposes, time is not a universal absolute but a social convention. The progress of time into the future is steady and inescapable, but "what time it is" *right now* is something we're free to choose. As we saw with the difference between the Gregorian and Mayan calendars, there is no single inevitable system, and the particular system we choose says something about our values and priorities.

Given the social aspect of time, the story of the establishment of our modern system is necessarily one of consensus building and negotiation. The US portion of that story also unfolds in a quintessentially American fashion: the process was set in motion by a tiny number of highly motivated individuals and brought to completion through political lobbying by massive corporations.

RAILROAD TIME

The railroad network in the United States, like many other things in the country's history, grew out of a patchwork of privately owned local operations. Individual companies sprang up in cities all over the country and began laying track and running trains in their immediate area, slowly extending their networks to reach other nearby cities. As the rail networks began to overlap, neighboring companies sometimes merged with one another, but much of the time they simply created junctions between lines where passengers would switch from one system of trains to another for the next leg of their journey.

Contrary to some tellings of the story, time standardization did not happen because this multiplicity of networks created problems or crashes.* Railroad companies realized the benefits of time standardization within their own networks very early, and began adopting a single standard time to simplify their internal scheduling as early as the mid-1830s. With each line controlling its own tracks and trains, there was no need for universal standardization across the entire network. Internal scheduling problems were easily fixed by adopting a single time for a given company's lines, but it didn't need to be any *particular* time, and for any given company the number of transfer points was small enough that schedules could be sorted out by direct negotiation between companies.

So, while there was some push toward consolidation—for example, in 1849 all the rail companies serving New England agreed to adopt a single standard time across their whole network—for the most part, rail companies were content with a patchwork system of different "railroad times" for each of the different systems. In the early 1880s there were over 300 individual railway lines operating in the United States, but 199 of those lines took their time from one of just seven cities (New York; Chicago; Philadelphia; Boston; Washington, DC; Columbus, Ohio; and Jefferson City, Missouri); the remaining lines took their time from among 50 other cities. The system could be inconvenient for travelers, who in some cities might find multiple clocks showing

---

* The handful of nineteenth-century train disasters attributable to poor timekeeping were caused by issues within a particular network: stationmasters or conductors whose watches showed the incorrect time, and thus put two trains on the same track in opposite directions.

the different times kept by different lines—as many as six in a major hub like Pittsburgh—but it worked well enough for the rail operators. Charles Dowd, the principal of a women's seminary in Saratoga Springs, New York, proposed a system of time zones in 1869 very much like the one eventually adopted but was never able to get much traction with the railroads, because they simply did not see this as a problem requiring much effort.

The event that finally set time standardization in motion was astronomical in nature, a dramatic display of the aurora borealis. At the time, the nature of the aurora was not well understood—we now know it to be caused by high-energy particles from space striking atoms in the upper atmosphere—and the astronomer Cleveland Abbe hoped to get some insight from the spectacular light show seen across most of the northern United States in April of 1874. As the chief meteorologist of the recently created United States Weather Bureau within the US Signal Service, Abbe had access to a national network of trained weather observers, who monitored atmospheric conditions in their local areas and sent data by telegraph to Abbe's office in Washington, DC. Around 80 of these observers recorded data on the aurora on the same evening, but Abbe's efforts to combine their distributed observations of the aurora and correlate them with other phenomena were frustrated by the patchwork of local times they had used to record what they saw.

Annoyed by this lack of uniformity, Abbe did two things. First, he reformed the instructions sent to his weather observers, ordering them to record their observations only according to Washington, DC, time, distributed by telegraph from the US Naval Observatory. He also sent a letter to the American Metrological Society urging them to begin advocating for standardizing time across the nation. This society, a group of scientists and academics based at Columbia University, responded in the inevitable manner of such a group: they created a Committee on Standard Time to study the problem and named Abbe its chair.

While in many contexts such an action would have doomed the idea, as a civil servant Abbe was no stranger to navigating the ways of bureaucratic organizations, and he enlisted the help of two men who could actually make standardization happen in North America. In the United States, the key figure was William Allen, the secretary of the General Time Convention of railroad officials, while to the north it was Sandford Fleming, a Scottish Canadian engineer who was then the chief engineer of the Canadian Pacific Railway. Allen's

General Time Convention was the rail-industry body responsible for coordinating schedules among lines, and he was also the editor of the *Traveler's Official Railroad Guide of the United States and Canada*. This combination gave him both inside access to deliberations among lines and a public platform from which to advocate for changes of time.

Fleming, the architect of much of Canada's rail system and an enormously respected figure, had already become convinced of the need for standardized time and in the 1870s started publishing pamphlets about the issue and lobbying governments and scientific societies. Befitting a citizen of the British Empire near the peak of its power, he was thinking on a global scale and proposed a system whereby every clock in the world would show the same time, based on the position of the sun at that moment. Fleming assigned letters to lines of longitude spaced by 15 degrees, starting from Greenwich, and under his scheme when the sun was at its maximum height on a particular meridian—C, say—it would be C o'clock everywhere in the world. While Abbe didn't sign on to this scheme, he and Fleming began corresponding and coordinating their efforts around 1880.

While Allen was technically invited to join the Metrological Society's Committee on Standard Time in 1879, he only learned of the invitation in late 1881. Once he heard about the efforts of Abbe and Fleming, he recognized the benefits of standardizing time, but being a railroad man to the core, he decided to reshape the system in a way that better suited railroad interests. As he drily put it in an editorial aimed at fellow railroaders, this was not a matter to be left to "the infinite wisdom of the state legislatures," and action at the federal level was no better: "there is little likelihood of any law being adopted in Washington, effecting [sic] railways, that would be as universally acceptable to the railway companies."

With an eye toward preempting legislation, Allen produced a map dividing the United States into four time zones, with boundaries drawn to coincide with junctions between rail systems. These are nearly identical to what Dowd had proposed 10 years earlier, and with a few exceptions are fairly close to the state-border-based time zones used today. Allen then used his influence within the railroad world to cajole and coerce individual railroads to sign on to his plan. The plan was officially presented to the General Time Convention on April 11, 1883, and the final vote in favor (with each railroad getting a share of votes proportional to the number of miles of track it controlled) was announced on October 11 of that year: 27,781 miles in favor, 1,714 miles opposed. Including railroads not directly members of the convention brought the "pro" vote to an overwhelming 79,041 miles. Railroads across the country committed to adopting Allen's four-zone scheme, with the change to take effect on November 11, 1883. By the end of the following year, the only holdouts were two small railroads near Pittsburgh, and they finally gave in by 1887.

Map of railroad time zones, with pre-standardization time. From Carlton J. Corliss's *The Day of Two Noons* (Association of American Railroads, 1952).

In parallel with the effort to bring railroads on board, Abbe and Fleming continued to lobby local governments to sign on, now aided by the considerable political clout of the railroads. Various city and state governments agreed to align their time with that of the railroads, largely for the benefit of convenience: it was easier by far to shift the official local time by some minutes one way or the other than to keep track of both local and railroad times. As Allen later wrote, when he heard New York City's clocks ring noon for the second time (four

TELEGRAPHIC TIME

In order to establish and maintain time zones covering large stretches of territory, of course, we need a way to synchronize clocks that are widely separated. No human-built clock is perfect, and even the very best mechanical clocks, set to the same time and allowed to run independently, will inevitably slowly drift apart. If we want all the clocks in a given time zone to show precisely the same time, we need a way to check them against some reference clock and reset them to correct for that drift.

This is a familiar problem in modern life, where clocks are all around us and frequently need adjusting. The dashboard clock in my car, for example, tends to run a bit fast, and every few months I need to roll it back a couple of minutes; the pendulum clock in our dining room, on the other hand, tends to run slightly slow, so every now and then we need to nudge the hands forward so the chimes better coincide with the actual hours. In every case, the solution is the same: we check the time on a clock that we trust more (usually a smartphone getting the official time from the internet), and then set the less-trustworthy clock to match it.

There are long-standing methods in place to do this synchronization in a local manner: the bells on church or city clocks are the oldest and crudest such method. To synchronize clocks with the precision needed for accurate longitude measurements, major port cities began to establish visual indicators. The Royal Observatory in Greenwich, for example, started using a "time ball" in 1833. This was a bright red sphere on a tall spire above the main observatory building; every day at 12:58 PM the ball was hoisted to the top of the spire, and at precisely 1:00 PM it dropped down. Navigators on ships at the London docks could watch the ball drop and adjust their watches accordingly, thereby carrying the official London time with them. Other major port cities adopted similar practices to ensure that ships headed out to sea were carrying the correct time.

For people and businesses without a clear line of sight to a major observatory, synchronizing clocks requires sending messages. On a small scale, the time could be hand-carried by a courier with a good watch. Well into the twentieth century, Ruth Belville ran a time service in London: once a week she would take her family's heirloom watch to the Royal Observatory to have its time checked and certified against the official clocks,* then she would travel around the city visiting her list of subscribers so they could adjust their clocks to match. The Belville time service had been started by her father in 1836 and only ended in 1940, when war conditions made it unsafe for an 86-year-old to continue selling time on foot.†

Belville's time service continued as long as it did because it was cheap and social—Ruth's clients appreciated the personal attention and chitchat that came with her weekly visits. The core idea of the technology that replaced her was already established in her father's day: electrical signals sent out from the observatory to indicate the correct time. Not long before Ruth retired, the observatory launched a telephone service, allowing anyone with a phone to call at any time and hear a recorded voice relay the correct time. Before that, though, there was a time signal on BBC radio, and before *that* telegraph-based time services.

While these systems lack the charm of a personal visit from a nice woman with a watch, they are the only practical method for synchronizing clocks on the scale of modern time zones. Beginning in the 1850s, time-selling services began to operate out of major observatories, both national ones like the Royal Observatory in Greenwich or the US Naval Observatory in Washington, DC, and also prominent academic observatories like the one at Harvard University. Astronomers would determine the time from their observations, then generate a time signal that was distributed over telegraph lines to their paying customers.

Clock synchronization was a lucrative business, and also had some scientific importance, particularly in the area of geodesy. The ability to distribute accurate time over huge distances allowed surveyors to measure longitude with unprecedented precision, resolving nagging discrepancies. In the mid-1860s, the best astronomical determinations of the longitude difference between Europe and North America disagreed with each other, and with measurements based on transporting high-quality chronometers back and forth across the Atlantic, by

---

* The watch was a John Arnold pocket chronometer originally made for the Duke of Sussex in 1794, and presumably came into her father's hands via the Royal Observatory, where he worked. He had the original gold case of the watch replaced with silver, so as to be less attractive to thieves.
† Belville's watch was donated to the Worshipful Company of Clockmakers (one of the greatest names ever for a trade association) and displayed in their museum.

around four seconds. While that seems tiny compared to the issues that faced navigators in John Harrison's day, it was embarrassingly huge compared to the precision obtained on land using time signals sent by telegraph.

The discrepancy was resolved in 1867, using signals sent between New-foundland and Ireland on one of the first reliable transatlantic cables. As telegraph networks expanded across and between continents, other large and long-standing discrepancies in maps were gradually sorted out. The completion of a cable from the Azores in 1874, for example, settled once and for all the longitude of Brazil, whose uncertainty had previously been a whopping 30 seconds (a distance of around eight miles).

The large distance scales involved in these telegraphic operations, and the level of precision they demanded, required the development of a synchronization scheme that was to have enormous philosophical importance later (as we'll see in the next chapter). This system must account for, among other things, the speed of the signal passing between the locations of the two clocks; failing to do so will give a longitude difference that's either larger or smaller than the correct value, depending on the direction the time signal was initially sent.

To illustrate this, let's imagine two locations whose true time difference is one hour (15 degrees of longitude), connected by a communications system

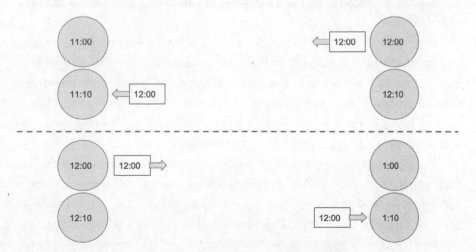

One-way signaling gives wrong answers for the longitude if you don't take travel time into account: a signal sent east to west measures a difference that's too small, while a time signal sent west to east leads to a difference that's too large.

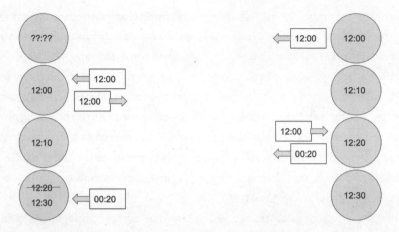

The two-way signaling method to synchronize clocks: a signal is sent east to west, then an acknowledgment is sent west to east. Afterward, half the round-trip time is added to the western clock to correct for the travel time.

that requires 10 minutes to transmit a signal from one to the other. If the cartographer at the eastern endpoint sends a prearranged signal at precisely noon (corresponding to 11:00 AM at the western station), it will arrive at 11:10 AM, and the two will conclude that the time difference between their locations is 50 minutes (corresponding to 12.5 degrees of longitude), a bit smaller than the true difference.

On the other hand, if we arrange for a second measurement but reverse the direction of the signal, with the western cartographer sending a signal precisely at noon, it will arrive in the east at 1:10 PM. This would correspond to a time difference of 70 minutes, slightly *larger* than the true difference. Hence, the two teams of cartographers would arrive at different answers for their separation, depending on which direction the time signal was sent.

The solution to this problem is to synchronize the clocks between endpoints first, by sending signals in *both* directions. The process begins with the eastern cartographer sending a signal precisely at noon, which arrives in the west at 12:10 PM on the eastern clock. Upon receipt of the signal, the western cartographer immediately sends a reply and resets her clock to noon. The reply will arrive in the east at 12:20 PM, giving the eastern cartographer the round-trip time; she can then send a message to her western counterpart with instructions to add half the round-trip time. This arrives at 12:20 PM on the western clock, which is corrected to 12:30 PM, after which both clocks will show precisely the same time. They can then determine their longitude difference by measuring the time of solar noon or the zenith crossing of some star at their two different locations.

This synchronization procedure is essential for nailing down longitude differences using telegraphic time signals. It also enables the precise synchronization of clocks across the large areas covered by the railroad time zones, allowing the kind of consistent and reliable timing of events across locations that Cleveland Abbe was after when he started to push for time standardization.

minutes after the first) on November 11, 1883: "Local time was abandoned, probably forever."*

GLOBAL TIME

William Allen's swift action preempted legislation that the North American railroads might not have found congenial, but lobbying by Abbe and Fleming and others had already set in motion governmental processes on a global scale. In the late 1800s, a pair of international conferences met to consider a number of questions relating to geodesy and the standardization of measurements, a set of questions that necessarily included standardization of time. Abbe and Fleming attended these as official representatives, and Allen testified as an expert on the standardization of railroad time.

The Seventh International Geodesic Conference met in Rome in 1883, for the first time with duly authorized representatives of all the major international powers, and among the most consequential issues they took up was the question of settling on a global standard for the prime meridian, the location designated as the zero of longitude. As we've noted several times before, this is a question that's closely connected with timekeeping, but also one with no clear scientific answer. The rotation of the earth gives us a timing-based method to determine *differences* in longitude, but there is no unambiguous landmark to determine an absolute longitude.

The choice of a starting point for longitude is thus a matter of convention, and as always with such questions, issues of national pride quickly came into play. Any astronomical observatory can establish its own standard of longitude based on timed observations at its particular location. While this would, in principle, allow for the prime meridian to be absolutely anywhere, there was a strong practical argument to be made for running it through an already-existing world-class observatory. The three most significant contenders for such an

---

* This was not quite true; some cities held out for several years, and a few that had signed on in 1883 later went back. Resistance to Allen's zone scheme was strongest near the eastern and western extremes of the zones—Maine, Ohio, and Georgia were among the most prominent dissenters—where the differences between clock time and solar time were most extreme. On the whole, though, the transition was remarkably smooth.

honor would be the Royal Observatory in Greenwich, the Paris Observatory, or (a distant third) the US Naval Observatory in Washington, DC.

All three of these had a solid claim based in scientific expertise, and the Paris Observatory had some additional argument from history and tradition, thanks to its sterling record dating back to the days of Jean-Dominique Cassini (whom we met in Chapter 6). In the end, though, as with most political questions, the ultimate resolution was rooted in practicality: in 1883 the British Empire was near the peak of its powers and dominated global trade. Something like three-quarters of the merchant shipping in the world at that time used maps and navigation tables originating in Britain that put the prime meridian in Greenwich. While several alternatives were proposed, in the end, avoiding the expense and inconvenience of reconfiguring such a volume of shipping won out.

Along with the setting of a zero line of longitude came a system of global time, calling for dividing the globe into time zones roughly 15 degrees of longitude in width (modified to respect political boundaries where necessary). Each zone would take a standard time based on the time in Greenwich, offset by some number of hours. The civil day was agreed to start at midnight, with the date line placed at the meridian opposite Greenwich (180 degrees east or west longitude), which conveniently falls in the middle of the Pacific Ocean.

The key results were agreed on in principle at the 1883 conference, then revisited at the International Meridian Conference of 1884, held in October in Washington, DC, at the invitation of President Chester Arthur (one of the results of Abbe's lobbying efforts).* Abbe, Allen, and Fleming again attended as official representatives. By the end of the month-long meeting, the recommendations from the previous year were adopted nearly unanimously (France and Brazil abstained in the final vote), and Greenwich became the official zero-longitude line for the world.

While the French contested the proposal to locate the prime meridian in the capital of their traditional rival, the US delegation readily went along with the British origin. The lack of a stronger push for an alternative can be traced, in part, to another savvy political move by Allen. The zones for standard railroad time established by Allen in 1883 were *already* linked to Greenwich time. In the discussions leading up to the adoption of his plan, Allen attributed this to

---

* I'm obliged to note that Arthur was an 1848 graduate of Union College, where I now teach.

a happy coincidence: when he drew up zones based on the junctions between railway lines, the central meridian of the Eastern zone turned out to be within a few seconds of 75 degrees west of Greenwich. Thus, he fixed the time difference for that zone at exactly five hours behind Greenwich time, with the other zones shifted by an hour each.

This choice, while apparently contingent on the specific locations of the zone boundaries (between the original drafting and final adoption, a change in operators shifted Atlanta into the Eastern zone, which ought to have changed the calculation slightly), was a convenient one for a couple of other reasons. For one, it synchronized railroad time with nautical time, since, as noted above, most global shipping was already operating based on a Greenwich meridian. It also avoided using the local time of any particular American city as the basis for the national system, which might have raised issues of local pride. The use of Greenwich time as a basis meant that *all* of the lines signing on to Allen's system had to change their clocks, but none of them had to change by much. And it set the stage for a smooth and easy transition to Greenwich time as the global standard a few years later.

SPRING FORWARD, FALL BACK

The railroad-based system of time zones implemented in 1883 by Allen and his colleagues remained in place in the United States, with minor modifications, until 1918 when Congress passed "An Act to preserve daylight and provide standard time for the United States," which established standard time zones for the country with more or less their current boundaries. As the name indicates, though, this also introduced another idea that's now accepted as routine: Daylight Saving Time (DST).

The "spring forward, fall back" ritual of resetting clocks by an hour twice a year began as a wartime energy conservation measure in Europe during World War I. The idea was that by shifting the official time forward an hour during the summer months, there would be less cause to burn critically needed fuel to generate electricity in the evening hours. After the end of the war, the practice of changing clocks was intermittently observed for many years in the United States, with different states and regions making their own decisions about whether to participate, leading to a weird patchwork of times that changed with the seasons. In 1966, the Uniform Time Act required the implementation of DST on a statewide basis, with states allowed to opt in or out subject to the approval of the federal government. As of 2020, all US states except Arizona and Hawaii observe DST in some form.*

If the initial introduction of standardized time zones was not sufficient evidence to prove that time is a matter of social convention, the DST system should remove any lingering doubt. While time zones at least have some basis in astronomy, given the different rising and setting times for the sun at different longitudes, DST is *purely* a matter of convention.

DST was created and continues to this day because it serves political and social ends. It nominally reduces energy use during the summer months, but mostly people just like having later sunsets during the summer. There are occasional proposals (including one in 2021) to move the clocks forward an hour and just leave them there, but these tend to underrate the unpopularity of the late sun*rise* in winter: under year-round DST, the sun would not come up until almost 8:30 AM in winter at the latitude where I live, which would require children to go to school and workers to offices in darkness. An experiment with year-round DST in the United States in 1974 was abandoned after six months, before even getting to the really dark months. As much as people grumble about the time changes, they like getting up before the sun even less.[†]

---

* In Arizona, the argument against DST is that an extra hour of daylight would *increase* energy usage by increasing demand for air conditioning. Hawaii is at a tropical latitude, where there is little seasonal change in the length of the day, so changing time would create a problem where none exists. Of course, it wouldn't be very American to have a stable consensus on any issue, so as of 2020 there are active legislative attempts to move off the DST system in Nevada, Florida, California, Washington, Oregon, and Tennessee.

† My own half-joking proposal is for an asymmetric time shift: have the clocks fall back an hour on a Saturday night in early November, then move them forward five minutes every Saturday starting after the December solstice, until they reach the standard time position in mid-March. This would give us late-summer sunsets, early winter sunrises, and an extra hour of sleep one weekend each fall, but avoid the disruption of jumping forward an hour in the spring. This illustrates why I will never hold elected office.

These days, we take the system of standard time zones and occasional arbitrary one-hour shifts for granted, and it's certainly more convenient for our highly mobile and globally connected society. We should not forget, though, what a dramatic change this represents philosophically speaking: the dawn of global telecommunications and rapid travel pushed us toward a more malleable, conventional picture of time defined in ways that work to make life easier.

The initial impact of this shift was, for most people, relatively minor—a changing of the clock by a few minutes in one direction or another, and the removal of the need to track railroad time and civil time separately. The rate of time continued to be a universal constant, though: while the precise time shown on a clock may be different in different places, two events separated by an hour in one place would be separated by an hour *everywhere*.

For philosophers and scientists, though, the implementation of conventional civil time zones foreshadowed a much more profound change in views: the abandonment of the entire idea of time as a universal absolute. And just a few years into the twentieth century, scientists realized that different observers will inevitably disagree not only about what time to set their clocks to, but the *rate* at which time passes, a surprising result that emerged directly from new developments in physics and philosophy.

Chapter 12

THE MEASURE OF SPACE-TIME

I n many popular tellings, physics in the late 1800s was in a state of arrogant complacency: between Newton's laws and Maxwell's equations,* plus the rapidly developing science of thermodynamics and energy, physicists thought they had everything figured out. To illustrate this attitude, many authors cite versions of a quote often attributed to the eminent British physicist Lord Kelvin in 1900:

> *There is nothing new to be discovered in physics; all that remains is more and more precise measurement.*

There is, however, no direct evidence that Lord Kelvin ever said any such thing. The closest solid citation is a lecture by his contemporary Albert Abraham Michelson, published in 1903, where he wrote: "The more important fundamental laws and facts of physical science have all been discovered, and these are so firmly established that the possibility of their ever being supplanted in consequence of new discoveries is exceedingly remote," and later attributed to

---

* We will discuss the important work of physicist and mathematician James Clerk Maxwell later in this chapter.

unnamed others the sentiment that "our future discoveries must be looked for in the sixth place of decimals."*

The fuller context of Michelson's lecture, however, made clear that this was not a dismissive statement of arrogance, but an affirmative statement regarding the importance of high-precision measurement. The material between those two quotes included an explicit statement that "it has been found that there are apparent exceptions to most of these laws, and this is particularly true when the observations are pushed to a limit, i.e., whenever the circumstances of experiment are such that extreme cases can be examined," followed by a partial listing of great discoveries made by observing small anomalies at extremes of precision.[†]

Michelson's invocation of "the sixth place of decimals" was actually part of a pitch for building up the capability to do measurements at that level. This should come as no surprise, because such measurements are the source of his own scientific fame: he was renowned as an exceedingly careful experimentalist who had made one of the best measurements of the speed of light in 1879. And as we'll soon see, another of his experiments at the very limits of measurement precision played a pivotal role in launching a revolutionary reexamining of the nature of space and time.

As for Lord Kelvin, who is regularly miscited as believing that physics was essentially complete, when given the opportunity to present a sweeping overview of the state of physics in 1900, the result was a lecture titled "Nineteenth century clouds over the dynamical theory of heat and light." In this he identified two major issues that he worried were obscuring "the beauty and clearness of the dynamical theory, which asserts heat and light to be modes of motion." One of these was the connection between heat energy and the light emitted by atoms, which was one of the problems that led to the development of quantum mechanics (which we'll talk more about in the next chapter). The other had to do with the nature and transmission of light, and as we'll see in this chapter, this eventually led to what we now call the theory of relativity.

In fact, Lord Kelvin and the other leading physicists of his day were keenly aware of the problems facing their field, which would lead to the creation of

---

* This appears, among other places, in a series of Michelson's lectures that was published by the University of Chicago Press as *Light Waves and Their Uses* in 1903; the full text can be found online.

† This list includes Ole Rømer's determination of the finite speed of light, discussed back in Chapter 8.

two theories that would radically reshape our understanding of the universe and how it operates. And each of these revolutions, in its own way, is rooted in an intensely practical question of timekeeping: how to keep clocks in distant places showing the same time.

THE PHILOSOPHY OF TIME

In 1898, the French polymath Jules Henri Poincaré published a philosophical essay called "La mesure du temps" ("The Measure of Time"), in which he considered a number of subtle issues concerning the perception and quantification of time. His analysis crystallizes the insight of the previous chapter: that time is a matter of convention.

Poincaré was trained as an engineer, at the École Polytechnique and then the École des Mines, and worked for a time as a mine inspector. His greatest talents lay in the field of mathematics, though, and he received a doctorate in the subject from the University of Paris in 1879 for research into new ways to study and solve differential equations such as those involved in Newton's laws when applied to the motion of planets in the solar system.

In 1887, King Oscar II of Sweden announced a mathematical contest, inviting submissions that would attempt to solve the "n-body problem" of multiple objects interacting with each other via Newtonian gravity. As we saw in Chapter 9, even the simplest version of this, the three-body problem involving the earth, moon, and sun is a formidable challenge, and expanding the number of bodies only makes it harder. Poincaré's entry to the contest brought to bear all the analytical tools he had worked with in the past and introduced some new ones, but in the end, the conclusion was negative: he showed that it is impossible to predict the future of such a system with absolute certainty.* While this is

---

* Interestingly, his initial submission reached the opposite conclusion: that an n-body system would always contain orbits guaranteed to be stable. This was judged to be the winner, but while it was being prepared for publication, a junior editor, Lars Edvard Phragmén, pointed out that Poincaré had missed what seemed like a minor point, leaving a gap in the proof. In attempting to clean this up, Poincaré discovered that the seemingly minor gap was, in fact, a major one. In the end, he was forced to completely reverse his conclusion. The printed copies of his original essay were hastily collected and destroyed, but the revised version *also* won him the prize.

somewhat ominous for the long-term future of the solar system, it was rightly hailed as a mathematical masterwork and won the prize; Poincaré's winning paper is now regarded as one of the founding documents in the mathematical study of chaos.

Having made a name for himself in academic science, Poincaré was named to the French Board of Longitude, the body responsible for coordinating the standardization of time in France and negotiating with the international community about the recently established system of time zones. In this role, he headed an unsuccessful attempt to decimalize time (i.e., to replace the traditional 12-hour system with one using 10 "hours" of slightly longer duration; this was eventually rejected as being too difficult to implement), and supervised the ongoing efforts to nail down the precise longitude of distant parts of the globe. These experiences undoubtedly informed Poincaré's thinking about time, and led in part to the publication of his essay.*

"The Measure of Time" did a thing that is characteristic of both great works of philosophy and late-night dorm room bull sessions, namely taking a close look at an issue that's normally taken for granted and finding hidden complexity in it. In this case, Poincaré looked at connecting *psychological* time, our qualitative sense of moving forward into the future, and *quantitative* time of the sort scientists like to measure. This may seem like something so basic and fundamental that it's only worth consideration by sleep-deprived college students (possibly with some chemical assistance), but in fact as Poincaré showed, a rigorous philosophical analysis makes clear that there is not an unambiguous connection between the two. Any system that purports to measure quantitative time in fact must ultimately rely on convention.

Poincaré considered two main issues: equality of duration and simultaneity of events. For the first, he asked how one could establish that two equal intervals of time were, in fact, equal: "When I say, from noon to one the same time passes

---

* The historian of science Peter Galison makes this argument at length in his book *Einstein's Clocks, Poincaré's Maps: Empires of Time* (Norton, 2004), which traces both men's thinking about the physics of relativity to their practical engagement with the science of timekeeping. It's an excellent book, and was highly influential in the development of my timekeeping course, which in turn became the book you're reading now.

as from two to three, what meaning has this affirmation?"* If you've read this far, of course, the answer is obvious in terms of the definition we've been using all through this book: we count the "ticks" of some system that we treat as a clock, and then compare the number.

As he pointed out, though, this tick counting requires an assumption that the "ticks" are equal, which isn't necessarily true. The ticking of a mechanical clock can be perturbed by vibration, friction, or changes of temperature; the flow rate of a water clock changes with the fill level and the temperature; even the apparent motion of astronomical objects depends on their complex orbits and the slowing rotation of the earth. We can and do correct for these effects, but we do so by using our knowledge of physics: we use Newton's laws to predict the motion of planets as well as to understand the perturbations acting on a mechanical clock. There's no guarantee that those laws of physics are true in an absolute and objective sense, though: physics-based corrections of time intervals necessarily involve the assumption that the laws of physics work in a particular way. Different laws might lead to different conclusions about the systems we use as clocks.

The second issue, the simultaneity of distant events, was posed by Poincaré as a grander version of a problem raised in a previous chapter: When we say that two events take place at the same time, what do we mean? Poincaré wrote:

> In 1572, Tycho Brahe noticed in the heavens a new star. An immense conflagration had happened in some far distant heavenly body; but it had happened long before; at least two hundred years were necessary for the light from that star to reach our earth. This conflagration therefore happened before the discovery of America. Well, when I say that; when, considering this gigantic phenomenon, which perhaps had no witness, since the satellites of that star were perhaps uninhabited, I say this phenomenon is anterior to the formation of the visual image of the isle of Española in the consciousness of Christopher Columbus, what do I mean?

---

* Quotes from Poincaré's essay are taken from the English translation by George Bruce Halsted that's available online at https://en.wikisource.org/wiki/The_Measure_of_Time (accessed June 11, 2021).

Again, the answer comes down to a matter of convention: We can say with confidence that the supernova Tycho saw in 1572 in fact took place many years earlier because we can use other means to establish the distance of the star that exploded.* Knowing that distance and the speed of light lets us determine how long the light took to get here from there and reliably date the original explosion.

But both of those measurements again presume a particular form of the laws of physics: we know the speed of light because of measurements like Ole Rømer's (from Chapter 8), which in turn rely on a discrepancy between the observed time of an eclipse and the time predicted using the physics of Kepler and Newton. Earth-based measurements similarly rely on a particular mathematical form for the laws governing the behavior of physical objects. And, of course, the idea that the speed of light is constant in both time and space is essential to both the measurement of that speed and its use to date the supernova. All of these are ultimately assumptions, chosen, as Poincaré noted, for reasons of convenience: the particular set of assumptions we make allow for Newton's laws to exist in the simplest mathematical form possible.

Like all good philosophy, this is heady stuff, involving a profound rethinking of the nature of time and our experience of it. It also connects very directly to practical issues in the measurement of time and longitude that Poincaré considered in his work with the Board of Longitude (as discussed in the previous chapter).

It's also no accident that the constancy of the speed of light is one of the key assumptions identified by Poincaré in his discussion of simultaneity. That question, whether the speed of light changes at different places or different times, was at the heart of one of the great crises that were then brewing in physics, which would lead to a revolutionary change at the dawn of the twentieth century.

THE SPEED AND NATURE OF LIGHT

Ole Rømer's observations of Jupiter's moons in the 1670s provided key evidence that light travels at finite speed, but the exact *nature* of light—that is, what sort

---

* The distance to the remnant of Tycho's supernova of 1572 is actually much greater than the lower bound Poincaré cited: modern measurements put it at least 8,000 light-years from Earth.

of thing was traveling from place to place—remained a topic of debate until the start of the 1800s. Isaac Newton's first major scientific work, describing his experiments on light and colors in 1672,* put forth a theory that light was a stream of microscopic "corpuscles," tiny particles with particular properties that correspond to the color, polarization, etc. of a particular light source.† On the other hand, Christiaan Huygens in 1678 put forth a theory of light as a wave, with different wavelengths and frequencies corresponding to the colors we see.‡ Newton's enormous reputation greatly boosted the corpuscular theory, which for many years offered the best explanation of polarization phenomena, while Leonhard Euler was an influential proponent of the wave theory in the middle 1700s.

The question was largely felt to be settled, in favor of waves, in the early 1800s, spurred by a surprisingly simple experiment by the English polymath Thomas Young. Young's original experiment used a thin "slip of card" inserted into a beam of light, but it's usually explained in terms of light passing through two narrow slits in some barrier and then projected onto a screen some distance away. If the slits are sufficiently narrow and close together, what you see on the screen is not a simple image of the two slits but an extended pattern of bright and dark spots.

The double-slit pattern discovered by Young is best explained as an example of the wave phenomenon known as "interference." When waves from two different sources are passing through the same medium, the pattern seen at any given point where they overlap depends on the phase of the individual oscillating waves, that is, where they are in their cycle of oscillation. If both waves are at a peak (or both at a valley) when they come together, the result is an extra-high peak (or extra-deep valley); this is referred to as "constructive interference." If,

---

* Later developed at greater length in his book *Opticks* in 1704. The 30-year delay between the two is partly attributable to Robert Hooke, who attacked Newton's 1672 work, the first of several nasty disputes between the two. Newton held off publishing further optical work until after Hooke's death in 1703.

† A particulate model of light had earlier been employed by René Descartes and Pierre Gassendi, though Newton made some key changes, most notably making properties like color intrinsic to the moving particles, rather than a transient property imparted to the medium.

‡ Some elements of a wave theory of light had also been used by the inescapable Hooke in his *Micrographia* in 1665.

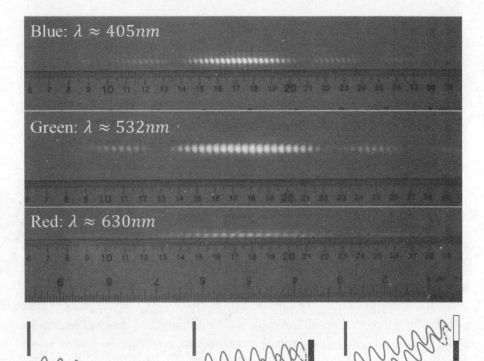

Double-slit interference. The top is the interference pattern for three different
wavelengths of light (increasing from top to bottom); the bottom shows how
the extra distance changes the relative phase of light from the two slits.

on the other hand, one of the two is at a peak while the other is at a valley, they
will cancel each other out ("destructive interference").

This is a familiar phenomenon with water waves—you can easily see it by
dropping two small rocks into a still pond and looking at the area between
them. As the two spreading circular wave patterns overlap, you will see strips
of water that appear basically flat separating regions where the waves are particularly prominent. We're not accustomed to seeing this happen with light,
because the wavelengths are extremely short—visible light ranges from vio-
let light at about 400 nanometers (0.0004 mm) to deep-red light at around

700 nanometers (0.0007 mm)—but this is what's happening in Young's experiment.

The extended pattern of spots seen by Young comes about because the relative phase of the waves depends on the distance traveled by the light. At a point directly between the two slits, waves coming from the top slit and waves coming from the bottom slit each travel precisely the same distance, so they will always arrive in phase, making a bright central spot. As you move to a point on the screen a bit away from the center, the light from one slit will need to travel a greater distance to reach that point than light from the other. As it travels that additional distance, it continues to oscillate and ends up with a slightly different phase at the screen, when it recombines with light that took the shorter path through the other slit. The extra path length increases as you move farther from the central point and will eventually add up to an extra half wavelength. When that happens, the two will be perfectly out of phase—one at a peak, the other a valley—and cancel out, leaving a dark spot.

If you keep moving out from the center, that extra path length continues to increase, and the difference in phases between the two waves continues to increase. Some distance farther from the center, the extra distance traveled increases to a full wavelength, so the wave taking the long route has gone through one full oscillation more than the one on the short route. This brings the waves back in phase—peak atop peak, valley below valley—and makes another bright spot. Moving still farther gets another dark spot (one and a half extra oscillations on the long route), then another bright spot (two full extra oscillations), and so on: the pattern will continue as far out as there's light to see.*

Young's results provided an enormous boost to the wave theory and helped inspire Augustin-Jean Fresnel to construct a mathematical theory of light as a transverse wave (with oscillations perpendicular to the direction of motion) that could explain essentially all of the optical phenomena then known, which seemed to settle the question of the nature of light. Fresnel's theory, submitted to a prize competition in 1818, was initially challenged on the grounds that it would predict a small bright spot in the center of the shadow of a circular

---

* The spreading of the light in the direction perpendicular to the slits is also a wave phenomenon, with the total extent determined by the width of the individual slits: the smaller the slits, the wider the pattern of bright and dark spots seen on the screen.

object; this seemed clearly nonsensical. When the existence of a bright spot in a round shadow was dramatically confirmed in experiments by the French scientist and politician François Arago, most physicists were convinced that light was a wave.

This left the question of *what* was waving; the solution to *that* came along in 1865 in the work of the Scottish physicist James Clerk Maxwell, who showed that light was electromagnetic in nature. Maxwell was adapting a concept first suggested by Michael Faraday, who thought about magnetism in terms of "lines of force" filling the space between objects (in modern terms, we call this a "field"). Faraday had excellent physical intuition, but as he came from a humble background, he didn't have the mathematical tools to turn his intuition into a formal theory. Maxwell had the formal education Faraday lacked, and thus was able to use the field concept to unify all known phenomena associated with electricity and magnetism.

In modern notation, the rules governing electromagnetic phenomena are written as "Maxwell's equations," four compact equations describing the properties of electric fields and magnetic fields. In some sense, these set the template for future theoretical physics as the pursuit of mathematically elegant theories that can be expressed in a form compact enough to put on a T-shirt; you may have seen them in this form, often paired with the phrase "Let There Be Light!"

The crucial elements for understanding light are a pair of equations connecting patterns of one field in space to changes of the other field in time. "Faraday's law" describes how a magnetic field that varies in time creates an electric field perpendicular to the magnetic field; this is the key principle that allows us to generate electricity by using steam or rushing water to spin loops of wire in large magnetic fields. Conversely, the "Ampère-Maxwell law" describes how an electric current or a changing electric field creates a magnetic field perpendicular to the electric field; this is the operating principle behind electromagnets. These two laws combined allow for the existence of an electromagnetic wave traveling through empty space: a changing magnetic field begets an electric field, that begets a magnetic field, and so on. The two fields travel along supporting each other, and they move at a very particular speed (commonly designated as c) that matches the known speed of light.

Maxwell's theory of electromagnetic waves was confirmed in the 1880s thanks to a series of ingenious experiments by the German physicist Heinrich

Hertz, using oscillating electric currents to generate waves at particular frequencies and demonstrate that their properties matched Maxwell's predictions. Displaying the business acumen of a great physicist, Hertz reportedly disclaimed this as merely a theoretical curiosity, saying that his results were "of no use whatsoever . . . this is just an experiment that proves Maestro Maxwell was right—we just have these mysterious electromagnetic waves that we cannot see with the naked eye."* Within a few years of his experiments, though, people were using pulses of Hertz's electromagnetic waves as a "wireless telegraph" to send messages across long distances; over time, this evolved into modern radio telecommunications.

Thus, by the late 1800s, when Poincaré was writing, the notion of light as an electromagnetic wave was well established. The one lingering issue concerned the *medium* in which light traveled. The other waves we encounter in everyday life are disturbances in some medium—water for ocean waves, air for sound waves, etc.—and as a result, from the beginning, light waves were presumed to travel in a medium of their own, dubbed the "luminiferous aether."† This was believed to be a pervasive medium that filled the universe and played host to the electric and magnetic fields that constitute light waves.

The properties required for this aether, though, are kind of problematic. Certain observations seem to require the aether to be nearly massless (so that it doesn't perturb the motion of the earth and other planets), while others seem to require it to be *infinitely* massive (to explain the lack of apparent forces when light is absorbed and emitted). These improbable properties were among the difficulties that Lord Kelvin and others noted regarding the physics of light waves. This sort of contradiction would seem to call into question whether the aether really exists at all, but the idea of a wave without a medium to support it was extremely troubling on a philosophical level, so physicists fought to keep it. Some began to look for direct evidence of the aether, to settle the question once and for all.

The key to hunting the aether is the single speed c that emerges from Maxwell's equations, the characteristic speed predicted for electromagnetic waves.

---

* This is repeated in lots of places, most of which seem to trace back to the textbook *Dynamic Fields and Waves*, edited by Andrew Norton. CRC Press, 2000.
† Historically this was spelled both "ether" and "aether" by different authors; I prefer the latter, as it has a nice prog-rock sort of quality.

This was presumed to be the speed of light relative to the aether, that is, the speed that would be measured by an observer at rest floating in the aether. In that interpretation, there ought to be some variation in the speed of light measured by an observer moving through the aether, in much the same way that a person riding on a bike will perceive a strong wind in their face when they're riding into what a person standing beside the bike path feels as only a gentle breeze. The speed of the observer moving through the aether should be added to the speed of light, in the same way that the speed of the bike is added to the speed of the wind.*

Of course, the speed of light is enormous—186,000 miles per second in American units, 300,000 km/s in the rest of the world—so detecting a change in its speed requires both a fast-moving source and an extremely sensitive measuring technique. The fast motion came from the earth itself, which orbits the sun at a speed of about 30 km/s, faster than any human-made object in history, and the sensitive technique was the brainchild of a Polish American physicist with a passion for precision measurement.

The Most Famous Failed Experiment in History

Albert Abraham Michelson (whom we met singing the praises of precision measurement at the start of this chapter) was born in what is now Poland, but his parents emigrated to the United States when he was two years old, to work as merchants in the mining towns of California and Nevada. Albert was granted admission to the US Naval Academy by a special dispensation from President Ulysses Grant.† He excelled in sciences, returning as an instructor after a brief period of service at sea. He was particularly fascinated by the behavior of light, and while at the academy in 1879 made one of the best measurements of its speed to that time, within a tenth of a percent of the modern value.

---

* Or subtracted, if you're biking downwind or moving away from the light source.
† Michelson was unable to secure an appointment to the academy from his own congressmen, so he traveled across the country to make a personal appeal to the president. As it turned out, Grant had already made all the academy appointments he was normally allotted, so a dejected Michelson headed home. He was intercepted at the train station, though, by a messenger with the news that Grant had decided to appoint him anyway (and throw in appointments for children of some political allies while he was at it). In later years, Michelson would joke that his career had been started by an illegal act.

He left the navy in 1883 for a position at the Case School of Applied Science in Cleveland, where he carried out the experiment that is his most famous contribution to physics: a search for the elusive aether. His experiment was based around what is now called a "Michelson interferometer," a device that uses the interference of light waves to make incredibly sensitive measurements.

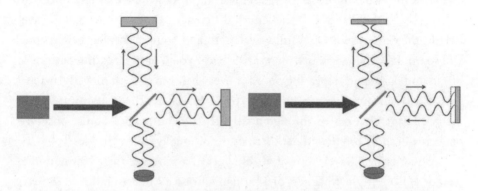

A Michelson interferometer. Light coming from a source at left hits a beam splitter, with half the light reflected up and the other half transmitted. Both paths go to mirrors and are reflected straight back to the beam splitter where they are recombined. If the two arms are equal length (left), the recombined waves are in phase and interfere constructively at the detector. If one mirror is moved back a quarter of a wavelength (right), the recombined waves are out of phase and interfere destructively.

The center of a Michelson interferometer is a beam splitter that reflects 50 percent of the light that falls on it, while allowing the other 50 percent to pass straight through.* This is aligned so it sends the reflected and transmitted beams on two perpendicular "arms," each leading out to a mirror that reflects each beam straight back to the beam splitter. On the return to the beam splitter, half of the originally transmitted beam is reflected and half of the originally reflected beam is transmitted, and these two new beams overlap and fall on a detector.

What you see at the detector is an interference pattern that depends on the phase of the beams from the two different arms. If paths traveled by light going down and back the two arms are precisely the same length, and the

---

* In Michelson's day this was just a bad mirror, a piece of glass coated with half as much silver as usual. More modern versions use very thin coatings applied to glass plates.

interferometer is stationary, both beams have traveled the same distance, so like the beams reaching the center of a double-slit pattern, they arrive perfectly in phase and interfere constructively to produce a bright spot. If one arm of the interferometer is longer than the other by one-quarter of the wavelength of the light used, the beam sent down that arm has to travel an extra half wavelength, and thus the two will be out of phase, leading to destructive interference and a dark spot at the detector. If the length difference is increased to half a wavelength, the extra distance is a full wavelength and you get another bright spot, and so on. If you slowly move one mirror back, you'll see a repeating pattern of bright and dark spots, repeating once for every half wavelength moved. In a real device with an input light beam that covers some size, the actual output will include beams that passed through at slightly different angles, so the observed pattern will show bright and dark stripes, traditionally called "fringes."

A quarter of a wavelength of visible light is a bit more than 100 nanometers, or 1/1,000th the thickness of a human hair, so a Michelson interferometer offers exceptional position sensitivity. These days, they are used in a wide variety of precision positioning systems, making computer chips and other devices requiring nanometer precision. On a grander scale, the Laser Interferometer Gravitational-Wave Observatory (LIGO) is a pair of absolutely enormous Michelson interferometers, with arms several kilometers in length, used to look for the tiny stretching and compression of space caused by gravitational waves passing through the earth.*

Michelson developed this device not as a technology for position measurement, though, but a tool for hunting the aether. If that's the goal, you fix the length of the arms (i.e., the distance between the beam splitter and the mirrors) and *move* the interferometer along one of the arms. Light headed in the opposite direction as the interferometer effectively moves at a slightly higher speed: the usual speed of light plus the speed of the interferometer. Light traveling in the same direction as the interferometer, meanwhile, seems to be going slightly

* With its two installations in Hanford, Washington, and Livingston, Louisiana, operating as a single observatory, LIGO is remarkably successful at this, detecting the gravitational waves from a colliding pair of black holes a billion light-years away almost as soon as the interferometers were switched on in 2015. The 2017 Nobel Prize in Physics went to Barry Barish, Kip Thorne, and Rainer Weiss for the development of LIGO, one of the most well-deserved prizes in recent memory.

slower: the usual speed of light minus the speed of the interferometer. These don't cancel each other out, so the round-trip time for light going out and back an arm aligned along the direction of motion is longer than it would be for a stationary interferometer.

The effective speed in the direction perpendicular to the motion would also change, but by a smaller amount, so the two arms have slightly different round-trip times. This difference in speed along the two arms would thus have the same effect as a change in length: light from one arm would be slightly delayed relative to the other, leading to a change in the pattern of bright and dark spots.

It's not possible to stop the earth's motion while you align an interferometer and then set the planet back in motion, of course, but Michelson realized that it's enough to rotate the interferometer. As first one arm, then the other is aligned along the direction of the earth's motion through the aether, they will alternate being "fast" and "slow," and at some intermediate point the effective speeds will be equal. This would be visible in the experiment as a back-and-forth wobble of the pattern of bright and dark fringes as you rotate it through a complete circle. The greater the speed difference, the bigger the wobble, so the extent of the shift gives a direct measurement of the speed of the earthbound interferometer relative to the aether.

Michelson tested the technique in 1881 with an interferometer that was too small to make a definitive measurement, but it showed that the basic idea was sound. On moving to Cleveland, he and his colleague Edward Morley constructed a much larger interferometer with additional mirrors to bounce the light back and forth multiple times, leading to an effective arm length of some 11 meters.* To minimize the effect of vibrations, they mounted this on a massive block of sandstone, and to allow smooth rotation of the block, they floated the whole thing on a vat of mercury.† They calculated that the fringes in their interferometer should shift by about 0.4 times the width of a fringe, and that they could reliably detect shifts as small as 0.01 times the width of a fringe. If

---

* Obligatory alumnus brag: Morley was a graduate of my alma mater, Williams College, class of 1860.
† Health and safety standards for academic research labs were very different in the 1880s.

the earth's motion through the aether changed the speed of light, Michelson and Morley were perfectly prepared to spot it.

With the apparatus constructed, the two scientists repeated their experiment over and over, rotating the sandstone block and running it at different times of day, so the rotation of the earth would change the alignment of the interferometer with the orbital motion, and watching the fringes through a telescope eyepiece to look for the signature wobble expected due to the "aether wind." They saw absolutely nothing. Being good and cautious experimenters, they reported in their paper of 1887 that the velocity of the earth relative to the aether was no more than one-sixth of its orbital velocity relative to the sun.

Morley would go on to repeat the experiment in 1902–05 with another colleague, Dayton Miller, using a much larger interferometer, but with the same null result. Miller carried the experiments on for many years afterward, and other experimenters adopted various refinements of the technique, but no conclusive evidence of a fringe shift has ever been seen. To the best of our ability to measure it, the speed of light does not change due to the motion of the earth. This presents a major problem for the entire theory of the luminiferous aether.

RELATIVITY BEFORE EINSTEIN

The Michelson-Morley experiment, together with some other ambiguous results, set physicists to seeking a way to explain the apparent constancy of the speed of light. These speculations form the backdrop for Poincaré's philosophical musings about the conventional nature of time.

One obvious possibility to account for the null result of Michelson and Morley is "aether drag," the idea that as the earth moves through space, it pulls some of the local aether along with it. This would reduce the apparent speed of the earth and its contents relative to the aether near the surface. The basic idea is a sort of reversal of the "wind chill" effect, that staple of weather reports at higher latitudes, where the wind pushes away the thin layer of air that's already been warmed by body heat, making a windy winter day feel colder than one with still air at the same temperature. The aether carrying the light through the interferometer would be moving at more or less the same speed as the interferometer, in the same direction, and thus would carry light at the same speed along both arms.

This was an active area of investigation for a time, and one of the later variants of the Michelson-Morley experiment was conducted outdoors atop Mount Wilson in California in order to better expose the apparatus to aether that should be moving. The results were no different, though. Aether drag was eventually ruled out as an explanation because the amount of drag needed to explain the experimental results should also produce a measurable aberration in the positions of stars depending on the earth's motion, but no such aberration was seen.

As an alternative to aether drag, some physicists began to speculate about the possibility of changes to the measurement apparatus that would disguise the "true" speed of light. The Irish physicist George FitzGerald suggested in 1889 that the result could be explained if moving objects shrink along the direction of motion through the aether. This would shorten the along-the-motion arm of the Michelson interferometer by exactly the amount needed to compensate for the increase in travel time due to motion through the aether. Such a shift would be undetectable, though, because *everything* would shrink along that direction, including any apparatus used to measure distances: if the ruler shrinks by the same amount as the object being measured, the apparent length remains unchanged.

At around the same time as FitzGerald, the Dutch physicist Hendrik Antoon Lorentz independently had a similar idea while working on a theory of the electron. From the mathematical structure of Maxwell's equations, Lorentz worked out a set of equations describing the changes that would be required for observers moving relative to the aether to measure a constant speed of light. This included not only shrinking along the direction of motion, but also a change in time; according to Lorentz, a moving observer would only measure a "local" time, which would "tick" at a slightly slower rate than the time measured by someone at rest with respect to the aether (which Lorentz viewed as the "true" universal time). This local time would also vary according to the *position* of the observer along the direction of motion: if we choose some moving observer as a reference point, their local time will lag behind that of observers in front of them moving at the same speed, but run ahead of that of observers behind them.

FitzGerald and Lorentz originally conceived of the shrinkage of objects as a real physical effect caused by motion through the aether changing the strength of the electromagnetic forces that hold objects together. As Lorentz continued to explore his ideas, though, it became clear that this couldn't work. When Poincaré began to look into this problem in the 1890s, he was the first to point out a crucial fact that's central to the modern understanding of relativity: Lorentz's local time emerges naturally when you attempt to synchronize a set of clocks that are moving.

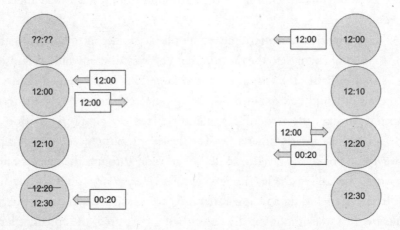

The two-way signaling method for synchronizing clocks is discussed in the previous chapter.

As Poincaré noted in "The Measure of Time," establishing that two events are simultaneous requires some way of establishing a common time between two locations. The most effective means of doing this is the clock synchronization scheme used by telegraphers in Poincaré's Bureau of Longitude (described in Chapter 11). To briefly recap: a time signal is sent from some central time-keeper to a surveyor at a remote location, and when the signal arrives, they immediately set their clock to the prescribed time, and send a return message. On receipt of the return message, the central authority sends out a message to adjust the clock by half the round-trip time; this fixes the error caused by the time needed for signals to travel from the center to the surveyor, and ensures that both clocks will show the same time. The properly synchronized clocks then allow the surveyor to make an extremely precise measurement of the longitude.

This method works, but it relies on the idea that the speed of the messages is the same in both directions. This seems to be empirically true, as verified by tests like the Michelson-Morley experiment. Poincaré pointed out, though, that things look very different from the standpoint of an observer sitting at rest in the aether watching the synchronization process proceed.

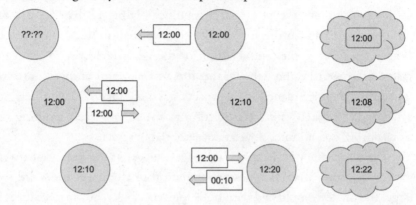

The two-way signaling method for synchronizing clocks gives incorrect results when employed by two surveyors who are moving, according to an observer at rest (right) with respect to the aether.

To make the effect more obvious, let's imagine that the speed of the planet carrying our surveyors through the aether is extremely high—around 3/10ths of the speed of the messages carrying their time signals. As they do the measurement, they're being watched by an observer floating in space, perfectly at rest with respect to the aether and equipped with perfectly synchronized clocks, also at rest in the aether, to establish the precise times at which signals are sent and received. This observer can then reconstruct the synchronization process in detail, and what they see looks very different from what the moving surveyors see.

If the initial message is sent out at precisely noon according to both the central timekeeping authority and our aetherial observer, the authority and the surveyor will see it take 10 minutes to cover the distance between them, just as in the earlier example. Since they're moving together, carried through the aether by the same planet, nothing will seem out of the ordinary to them. To the aetherial observer, though, the surveyor is moving *toward* the incoming message (and the authority away from it), and thus the distance the signal needs to travel is slightly shorter. The surveyor receives the message at a point slightly closer to where the authority was when the message was sent, and the true arrival time is 12:08 by the aetherial clock, not 12:10.

On the return trip, the aetherial observers see the central authority running *away* from the message, making the distance to be covered longer and increasing the travel time. These two effects do not cancel each other out—the increase in time on the return trip is, in fact, significantly bigger than the decrease on the outbound trip, so the round-trip is completed at 12:22 on the aetherial clock. According to the central authority, though, which sees no change in the distance covered by the light (again, the surveyor and authority are moving together with their planet), this process has only taken 20 minutes, not 22. Thus, they send a 10-minute correction to the survey team, resulting in both clocks being two minutes slow to the aetherial observer.

The discrepancy between the times gets bigger as the distance between the parties increases. If we double the separation so the travel times are twice as large, the process would end with both moving clocks showing 12:40, at 12:44 aetherial time. A network of clocks at different positions synchronized by this process, then, would look to our aetherial observers as if they were improperly synchronized, with the difference from true time getting worse the farther you get from the central clock. When you work through all the details carefully, the synchronization error seen by the aetherial observers exactly matches Lorentz's local time.

WHAT ABOUT EINSTEIN?

The mathematical equations at the heart of the theory of relativity were first derived by Hendrik Lorentz and, as a result, are known as the "Lorentz transformation,"* which had reached their modern form by 1904. The connection between Lorentz's local time and clock synchronization was first noted by Poincaré in 1900, and by early June 1905 he had a complete derivation of the

---

* Sometimes known as "Lorentz-FitzGerald." Of course, as is almost always true in history, this leaves some other people out: Joseph Larmor independently came up with a similar set of equations, and all of them were indebted to the work of Oliver Heaviside in reformulating Maxwell's equations.

TRAINS OF CLOCKS

It may not seem obvious that such mis-synchronized clocks would lead to dramatic changes in measurements of space and time, but in fact many of the odd features of relativity can be explained just through this sort of arrangement. In particular, comparisons of clocks whose synchronization is off in just the right way can lead observers in motion to the conclusion that moving clocks run slow, that moving objects shrink, and that nothing can move faster than a certain maximum speed. These are exactly the troubling results predicted by Lorentz and FitzGerald.

To see how to get changes in space and time from mis-synchronized clocks, let's consider an ingenious thought experiment devised by N. David Mermin in his book *It's About Time: Understanding Einstein's Relativity.** Let's imagine we have two trains running on parallel tracks in opposite directions, and each car contains an observer with a clock and a narrow window through which, at appropriate intervals, they can see the observers and clocks in the other train. Being both scientifically inclined and very bored, the observers on each train record what they see through the windows and attempt to use these measurements to deduce the properties of the other train.

Unbeknownst to the observers on the trains, though, their clocks have been adjusted to be out of synch in exactly the way that one would expect from the problem described above. Seen by an observer positioned next to the tracks as they pass, who can see *all* of the clocks on a given train, each clock is 2 seconds ahead of the clock in the car in front of it. So, if the clocks in the lead car of each train are perfectly synchronized as they pass at 0:00, the second car in each train shows 0:02, the third 0:04, and so on. Each train takes six seconds to cover the length of one car (according to an observer outside the trains), so the next time the cars align with each other, the lead cars show time 0:06, the second cars 0:08, and so on. This is the level of mis-synchronization one would expect for a train moving at one-sixth of a car length per second if the "speed of light" used in synchronizing the clocks were one-fourth of a car length per second.

---

* Now in its sixth printing, Mermin's book was originally published by Princeton University Press in 2005.

This is, admittedly, the sort of practical joke that would only occur to a theoretical physicist, but it has profound consequences for what the passengers in the cars conclude about the world from the observations they can make. When they try to draw conclusions about the other train—its speed and length—and the clocks it carries, they come up with answers that seem bizarre.

So, what conclusions would an observer in one train draw about the other train? Well, one easy measurement is to determine the relative speed of the trains, which we can do from looking at the recorded sightings of a single car. For example, the lead car (car 0) of the top train, which is seen directly opposite car 0 of the bottom train at 1:00 on both clocks, is opposite car 2 at 1:10 on the bottom-train clock, opposite car 4 at 1:20, and so on.

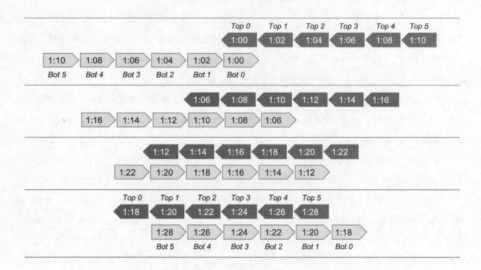

Four instants in the trains-of-clocks thought experiment showing how the mis-synchronized clocks on the two trains lead observers to conclude that the clocks on the other train are running slow and the other train has shrunk.

From these measurements, an observer on the bottom train would conclude that it takes 10 seconds for the other train to cover two car lengths, a relative speed of one-fifth of a car length per second. Right away, we notice something odd: the actual relative speed, according to our observer by the tracks, is one-third of a car length per second (each train is moving at one-sixth of a car length per second, in opposite directions, so their speeds should add), but the mis-synchronized clocks lead the observers on the bottom train to conclude that the top train's speed is slower.

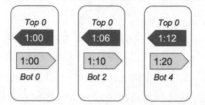

Three instants when car 0 of the top train is visible to observers on the bottom
train, leading them to conclude that the top train's clocks are running slow.

We can notice another issue in these images, though, namely that the clocks are ticking at different rates. When the two car 0s are aligned, both clocks show 1:00, but when the lead car of the top train is opposite the bottom train's car 2 at 1:10, its clock shows only 1:06. At 1:20 in bottom-train time, top car 0 is next to bottom car 4, but its clock shows only 1:12. The clock in the other train thus seems to be ticking too slowly by a factor of three-fifths: for every second that passes on the bottom train, 0.6 seconds pass on the top train. The moving clock on the top train is running slow, exactly as predicted from Lorentz's local time.

A final observation the bottom-train observers can make is to determine the length of a car on the other train. To do this, they need to find observations where two cars on the top train were spotted at the same instant. Looking at the collection of observations made on the bottom train, one such instant is 1:20, when top car 0 is opposite bottom car 4, and top car 5 is opposite bottom car 1. Five cars' worth of the top train thus fit within three cars' worth of the bottom train, meaning that each top-train car is only three-fifths of a bottom-train car. The moving cars have shrunk, exactly as predicted by FitzGerald and Lorentz.*

The truly remarkable thing about these conclusions is that they are perfectly symmetric. That is, the observers on the top train will see exactly the same pattern of results: bottom car 0 passes top car 2 at 1:10, with its clock showing 1:06, meaning that its relative speed is one-fifth of a car length per

---

* A really astute observer will note that the top train clocks show two different times at this "instant" for the bottom train, which is what ultimately gives the game away.

second, and its clock is ticking slow by a factor of three-fifths. They will also see bottom car 0 next to top car 4 at 1:20, when bottom car 5 is opposite top car 1, leading them to conclude that the *bottom* train has shrunk by a factor of three-fifths.

In the somewhat contrived hypothetical scenario of two trains, these odd results come about because the clocks on each train have deliberately been mis-synchronized in a particular pattern: they tick at the same rate, but *appear* to be running slow. As pointed out by Poincaré, though, this is *exactly* the situation that happens when we try to synchronize moving clocks in the real world. An array of clocks that appear perfectly synchronized to any one particular observer will be out of synch to an observer moving relative to them, while their (moving) synchronized array will seem out of synch to the first observer.

The train scenario dramatically illustrates that once you have a breakdown of synchronization, you automatically get the weird effects we associate with relativity. The impossibility of clock synchronization mixes time and space—to observers in motion, what time it is depends on where you are—and the rest follows from that. Each observer will say that the other's clocks are ticking too slowly, and each observer will say that the lengths and distances between moving objects have shrunk. This is not an illusion: there are no measurements that either of the observers can do that will show anything but that moving clocks run slow and moving objects shrink.

While this scenario seems to be contradictory in ways that can make your head hurt, the idea of synchronization also provides a way to make sense of it all. If we think carefully about what it means to synchronize measurements and how to achieve that, we can understand how to reconcile the different measurements in a way that gives each observer a self-consistent and coherent story about what happened.

Timing Is Everything: Reconciling Measurements

The key to understanding why moving objects shrink is to remember the fundamental meaning of measuring the length of an object, which we made use of in the trains-of-clocks example. Measuring length is not a single action, it's two simultaneous acts: recording the position of the front end of the object and the back end of the object *at the same time*. This is a familiar problem for anyone who's done carpentry or landscaping, which often requires two people: one to hold one end of the ruler or tape measure against one end of the object being measured (setting the location of zero), and the other to read off the position at the other end.

This situation becomes more complicated when the object to be measured is moving, particularly if the length is large enough and the measurement precision high enough so the time needed to signal from one end to the other becomes significant. In these cases, the synchronization of measurements also becomes significant, and that process is complicated by motion.

For simplicity, let's imagine that we're interested in measuring the length of a fast-moving train car and comparing the results from a measurement made inside the train to those made by an observer by the tracks. One slightly contrived way to do this would be to arrange a pair of devices, one at each end of the car, that on receiving a signal will instantly make a mark on the ground outside. To ensure that these fire simultaneously, Henri Poincaré on the train can set up a light source in the exact center of the car: the ray traveling to the front will take the same time as the ray traveling to the back, the signals will be received simultaneously, and the marks on the ground will record the positions of each end of the car at that precise instant.

To Hendrik Lorentz standing at rest watching the train pass, though, this scenario seems very different. The two rays of light leave the source and travel at the same speed toward the front and back of the train, but the *time* required for each to reach its destination is different. The light heading toward the back of the car has a shorter path to travel, because its detector is racing toward it. Meanwhile, the light headed toward the front of the car has to cover more ground, because its detector is running away.

Hendrik, then, sees Henri's attempt to measure the length of the car as not merely contrived but improperly done: the mark at the rear of the car is made first, the mark at the front of the car a moment later, and in between the car moves some distance. The length marked on the ground is not the true length of the car, because the measurements were not properly synchronized. The length of the car itself, its ends measured simultaneously by Hendrik next to the tracks, is shorter than the length measured by Henri on the inside.

Again, the remarkable thing about this explanation is that it is symmetric. As Poincaré was the first to note, a crucial fact about this scenario is that it depends only on the *relative* motion or lack thereof. Any pair of observers who are stationary with respect to one another will confidently assert that their clocks and measurements are properly synchronized, but they will disagree with the clocks and measurements made by any other pair of observers moving relative to them.

This notion of relativity lets us turn the scenario around: as a matter of mathematics, Henri on the train is perfectly entitled to imagine himself and his apparatus as being at rest, while Hendrik rushes past headed toward the rear of the train. Physicists do this all the time, in fact, because it is more convenient: we study the motion of everyday objects relative to the earth, meanwhile neglecting the earth's rotation, its orbit around the sun, the sun's orbit around the center of the Milky Way, and so on.

If Hendrik has set up a similar light-pulse system to synchronize his measurements of the ends of the train car, Henri will see it fail in an analogous manner to that seen earlier: the light ray headed to Hendrik's rear will arrive first, as its detector rushes up to meet it, while the ray headed to the front spends some extra time in transit. Since Hendrik is headed in the *opposite* direction from the train, though, this results in the front of the car being measured first, and the back a moment later. From Henri's perspective, then, Hendrik's too-short measurement is the result of an improperly synchronized measurement, with the back end of the car allowed to move closer to the recorded position of the front before it's measured.

If, rather than measuring Henri's car, Hendrik were set up to measure something he regarded as stationary—the platform of the station, for example—Henri would see this unfold in exactly the same manner as Hendrik saw the measurement of the car. To Henri on the train, the platform is moving, and the "rear" of the platform (toward the front of the train) gets measured first, with the front measured a moment later. According to Henri, the platform is shorter than the length measured by Hendrik, because Hendrik's measurements were improperly synchronized.

This tale of mis-synchronized measurements highlights one of the features of the theory that is hardest for people to wrap their heads around. Common sense would seem to say that the measurements should go in one direction only: if Hendrik thinks Henri's train is too short, then Henri should see Hendrik's platform as too long. In reality, though, they each see the other's object shrink: Hendrik sees a shorter train car, while Henri sees a shorter platform.

These seemingly contradictory measurements are resolved by the realization that, to paraphrase the title of Mermin's excellent book, it's all about time. The strange effects and seeming paradoxes of relativity flow from the central observation that observers who are moving relative to one another will disagree about the timing of events in a way that leads to separated clocks being out of synch with the other. Lorentz's transformation equations, with their contracted lengths and local time for each observer, are a way to formalize this observation in a mathematically rigorous way. Given a set of position and time measurements made by one observer, Lorentz's equations let you say with absolute certainty what positions and times will be measured by any observer moving relative to them.

equations in their modern form. Poincaré also noted the crucial role played by relative motion, leading to the modern name of the theory of relativity.

Given that much of the mathematical and conceptual work was already done by 1905, why is it that we associate the theory of relativity primarily with Albert Einstein? Einstein was at this point still an unknown patent clerk in Bern, Switzerland, having been unable to secure an academic job after completing his studies at ETH in Zürich. He was not an active part in the ongoing discussion of these theories until his seminal paper ("On the Electrodynamics of Moving Bodies"), which was submitted in late June of 1905 (after Poincaré's paper).

Much of the reason we associate relativity with Einstein has to do with work he did later (which we will discuss in Chapter 14), but there is one crucial difference in the theories of 1905 that lifts Einstein above Poincaré and Lorentz in modern thinking. What sets Einstein apart is the status of the aether.

Well past 1905, both Lorentz and Poincaré continued to think of the aether as a real thing, a privileged universal frame, and the lengths and times measured by earthbound observers in motion as merely "local" or "apparent" values. Einstein discarded the aether altogether, and in so doing was able to ground the whole of the theory in a single elegant philosophical principle.

For all its intimidating mathematical complexity, the core idea of the modern theory of relativity, as first presented by Einstein, is simple enough to fit on a bumper sticker:

The Laws of Physics Do Not Depend on How You're Moving

This "principle of relativity" dates back to the work of Galileo, who first articulated it in his *Dialogue Concerning the Two Chief World Systems*, where it's a key part of a counterargument against the claim that the earth must be stationary because if it were moving, we would feel it. Galileo notes that if you perform any of a long list of experiments inside the sealed cabin of a ship in a calm harbor, there is nothing in any of the results that will allow you to determine whether the ship is sitting at rest or moving with constant speed. It's critically important that the cabin be sealed—if you were on deck and could see objects on land, you could quickly deduce whether you were moving. From inside, though, you can only make *local* measurements: tracking the motion of objects relative to the walls of the cabin. Those local measurements are unchanged by the ship's motion: gravity doesn't pull objects down any faster when the ship is moving, and you can't jump farther toward the bow than toward the stern. Galileo used this thought experiment to argue by analogy that the motion of the orbiting and rotating Earth in the Copernican system would be similarly undetectable.*

Einstein elevated and expanded this idea, making it the foundational postulate of his relativity theory. In dry academic prose, this is: "The laws by which the states of physical systems undergo change are not affected, whether these changes of state be referred to the one or the other of two systems of co-ordinates in uniform translatory motion," but it boils down to what we wrote on the aforementioned bumper sticker. He followed this with a second postulate (quoted in Chapter 8), separately asserting that all observers see the same speed of light, but this is somewhat redundant. If Maxwell's electromagnetism is part of "the laws by which the states of physical systems undergo change," then the constant speed of light follows automatically. Maxwell's equations predict a single speed of light, therefore everybody sees the same speed of light.

In this view, the very concept of the "luminiferous aether" likewise becomes redundant. There is no need for a universal rest frame that defines a "true" time and length; relative motion is all we can ever measure, so it's all that matters. Time, in this view, becomes a personal thing, with the individual experience of time determined by one's motion. Contrary to some of the more hysterical misinterpretations of Einstein, though, this is not a free-for-all. Lorentz's transformation equations provide a rigorous set of rules by which to reconcile these different experiences based on the relative velocities of any two observers. We can always construct a coherent picture of the timing of events observed at

---

* This is not perfectly true: there are some dynamical effects that clearly prove that the earth is rotating, but these are subtle and rely on the fact that circular motion necessarily involves acceleration, as the direction is always changing. The most famous of these is Foucault's pendulum, a staple of science museums, where a pendulum swinging from a very good bearing will appear to rotate over a 24-hour period: in fact, the plane of the pendulum's swing is remaining fixed, and the earth is rotating beneath it. The principle of relativity articulated by Galileo in 1632 and expanded by Einstein in 1905 applies only to motion at constant velocity: an unchanging speed in an unchanging direction. Extending this to include accelerating motion leads to general relativity, which we'll discuss in Chapter 14.

different places, provided we know how the people making the measurements are moving relative to one another.

There's some irony in the fact that what places Einstein above Poincaré in the area of relativity is the latter's unwillingness to follow his philosophical musings from 1898 to their end. While Poincaré wrote eloquently and convincingly of time measurements as ultimately a matter of convention, in physics he and Lorentz held to the idea of a "real" aether for years after Einstein's introduction of relativity made that superfluous. A generation older than Einstein, they were less willing to cast aside the comforting notion of a true and absolute universal time.

While Poincaré and Lorentz couldn't take that last step, others in physics and mathematics enthusiastically embraced the new ideas. Among them was one of Einstein's former professors, Hermann Minkowski, who noted that the new theory implied a motion-dependent mixing of space and time, in which moving observers see different times at different positions. Two events that one observer sees as simultaneous, separated only in space, will appear to a moving observer to take place at two different times, but slightly closer together.

Minkowski found an elegant way of expressing this velocity-dependent mixing of space and time in terms of four-dimensional geometry, which he famously introduced to the Assembly of German Natural Scientists and Physicians in 1908 by saying:

The views of space and time which I wish to lay before you have sprung from the soil of experimental physics, and therein lies their strength. They are radical. Henceforth space by itself, and time by itself, are doomed to fade away into mere shadows, and only a kind of union of the two will preserve an independent reality. *

Einstein did not immediately embrace Minkowski's geometrical picture but over the next few years changed his mind. The four-dimensional geometry of space-time is the centerpiece of Einstein's theory of general relativity, an accomplishment which justly made him the most famous and recognizable scientist in the world. This remains our best theory of the physics of gravity, and

---

* From Minkowski's "Space and Time," originally published as *Raum und Zeit*, Jahresberichte der Deutschen Mathematiker-Vereinigung, 75–88 (1909).

as we'll see in Chapter 14, this has consequences for our understanding of space and time that go well beyond anything Minkowski imagined. Meanwhile, let's turn to the *other* great revolution in physics of the early 1900s, which has provided us with the ultra-high-precision clocks needed to confirm the predictions made by relativity.

Chapter 13

QUANTUM CLOCKS

The year 2016 was one of political turmoil and partisan rancor leading to electoral results that many people found surprising and upsetting, most notably the "Brexit" referendum in the UK, committing the country to leaving the European Union, and the election of Donald Trump as president of the United States. At the time, many people joked about how the year felt absolutely interminable, the longest year of their lives.*

As it turns out, there was an element of truth to this, particularly for anyone under the age of 44. The year 2016 was, in fact, the longest year on record since 1972, by one full second.† If you were watching the official world time as midnight approached in Greenwich on December 31, 2016, you would've seen the time tick from 23:59:59 to 23:59:60 before rolling over to 0:00:00 of January 1, 2017.

That extra "leap second" extending 2016 was added to keep the official time in synch with the astronomical "day" defined by the apparent motion of the sun. These have been added intermittently since their introduction in 1972, with either the last day of June or the last day of December getting a 23:59:60;

---

* They obviously had no idea what 2020 had in store . . .
† Strictly speaking, it's tied with 2012, 2008, and 22 other years for the honor of second-longest year of the modern era.

229

1972 itself holds the record for longest year ever, since leap seconds were added in *both* months.*

The insertion of leap seconds also reflects the ultimate decoupling of time from astronomy, in that the official definition of the second no longer has any relation to the rotation of the earth. Since 1967, the second has been defined as "the duration of 9,192,631,770 periods of the radiation corresponding to the transition between the two hyperfine levels of the ground state of the cesium-133 atom." Leap seconds are needed because the motion of the earth changes over time at a level that is readily detectable with modern timekeeping technology. Having fixed the definition of a second in terms of universal constants, we are obliged to temporarily change the definition of a "day" and a "year" by small amounts every now and then, to keep clock time and subjective time of day in synch with one another.

The story of *how* our clocks got to be this good is rooted in the most radical of the revolutionary changes in physics that occurred in the first half of the twentieth century, the development of quantum mechanics. Quantum theory is unquestionably one of the greatest intellectual achievements in human history, describing a fascinating world of wavelike particles and particle-like waves, and objects whose properties are fundamentally indeterminate. The infamous weirdness of quantum physics makes it all the more amazing that this strange picture of the universe led us to the construction of clocks of unparalleled precision.

LINES OF LIGHT AND DARK

There are many different entry points into the story of quantum physics, because the final form of the theory came about through the convergence of many lines of evidence from different areas of physics. The path that leads most directly to atomic clocks, though, in some sense begins with the collapse of a house in Munich.

---

* As it turns out, 1972 was also a nasty political year in the United States, with that fall's presidential election marked by the Watergate burglary and scandal that eventually led to the resignation of President Richard Nixon.

Joseph Fraunhofer was born in Straubing, then in the kingdom of Bavaria, in 1787, but was orphaned as a boy and found himself apprenticed to a glassmaker in Munich named Philipp Anton Weichelsberger. The life of an apprentice circa 1800 was not particularly pleasant anywhere, but Weichelsberger was notably harsh even for his time, giving his apprentices only menial tasks and tightly restricting their leisure activities, to the point of forbidding young Fraunhofer a lamp by which to read in his off hours. This didn't seem to bode well for Fraunhofer, but in a rather cinematic twist, Weichelsberger's house and workshop collapsed in 1801, with the boy inside. This was a spectacular enough event to attract quite a crowd, among them the future king of Bavaria, Maximilian I Joseph. When Fraunhofer was extracted from the wreckage more or less unharmed, Maximilian was so moved by this miracle that he provided the boy with a stipend and introduced him to Joseph von Utzschneider, a royal advisor who also ran a glassmaking operation at the Optical Institute in Benediktbeuern.

Utzschneider saw to Fraunhofer's education, then hired him on to assist with the making of lenses and other optical elements. Fraunhofer turned out to be an exceptionally talented glassmaker, and under his direction, the Optical Institute was soon turning out glasswork of unmatched quality. For years, scientists had struggled to deal with lenses and prisms made from glass containing bubbles and other imperfections, but Fraunhofer developed new glass recipes and manufacturing techniques that produced pure, clear glass on a large scale. Thanks largely to his work, Bavaria became the preeminent source of scientific glass in Europe, giving the kingdom an economic boost that earned him a knighthood, making him "von Fraunhofer."

One of the key steps in fabricating precision optical elements is the control of dispersion, the phenomenon where different colors of light bend by different amounts when entering a piece of glass. Dispersion is what creates the spread of rainbow colors from a beam passing through a prism, but it also leads to a blurring of images made by lenses, limiting the quality of telescopes. This "chromatic aberration" can be corrected by combining lenses of slightly different composition, so the extra bending of blue light in one piece is compensated by an extra bending of red light in the other.

Making high-quality "achromatic" lenses requires careful measurements of the bending of light at many different colors, and Fraunhofer set out to make the most exacting measurements attempted to that point. He set up a light source behind a screen with a narrow vertical slit, letting the resulting beam fall on the glass to be measured, and then used a surveying theodolite to make precise measurements of the angle of deflection for a given color of light. He was hampered in this effort, though, by a lack of stable references: the colors in a dispersed beam of white light blend smoothly into one another in a way that makes it difficult to ensure that you're always using the exact same piece of the spectrum when you measure the bending of "yellow" light.

Fraunhofer experimented with many different sources of light and eventually noticed that the spectrum of a lamp burning a mixture of alcohol and sulfur, passed through his apparatus using a high-quality prism, showed a bright orange line. That is, this particular flame produced much more light in one narrow range of wavelengths than elsewhere, producing a bright image of the vertical slit in that color. This was obviously a great help as a wavelength reference, but just one color, where he needed a wider range.

At this point, around 1815, Fraunhofer turned his apparatus toward the sun, hoping that this might reveal some similar bright features he could use. Instead, he found the opposite: "an infinite number of vertical lines, of different thicknesses. These are darker than the rest of the spectrum, some of them entirely black."* He assigned letters to the most prominent of these, some of which are still in use today—many physicists still refer to a particular feature in the yellow-orange region of the spectrum as "the sodium D lines."

In the course of his investigations, Fraunhofer invented an entirely new device for studying the spectrum of light, the diffraction grating. This came about when he was looking into Young's double-slit experiment (mentioned in the previous chapter), which also dispersed white light into its different colors. A true double slit passes very little light through it, so Fraunhofer added some additional slits, hoping to get a brighter pattern, and found that the interference pattern became sharper. This is another result of the wave nature of light.

---

* The presence of dark lines in the spectrum of sunlight had previously been noted by William Wollaston in 1802. As Wollaston's apparatus was greatly inferior to Fraunhofer's, he was only able to see a few of the most prominent lines. Other than a tentative attempt to identify these with the divisions between a small number of colors (the "ROYGBIV" spectrum we teach to school-age kids), they didn't attract much attention, and Fraunhofer was likely unaware of them when he observed more than 570 dark lines a decade later.

Expanding from two sources to many makes the condition for constructive interference more stringent: the bright spots of any particular color became narrower and the dark regions between them more distinct. As in the double-slit pattern, the constructive interference occurs when waves from neighboring slits travel paths that differ by a full wavelength of light. This means that longer wavelengths of light give patterns with a larger separation between spots, therefore, a grating also acts to disperse the colors of light.*

Fraunhofer built diffraction gratings with many slits starting with arrays of thin wires laid across finely threaded screws; he later replaced these with fine lines etched into glass surfaces. Light shined through these gratings was dispersed into its component colors in a manner similar to that achieved with a prism—only better controlled, as the dispersion depended only on the carefully manufactured spacing between slits instead of the difficult-to-measure properties of the glass itself. This dispersion let Fraunhofer measure the absolute wavelengths of various colors of light, as well as the light associated with the dark lines in the solar spectrum. He also used his newly invented grating spectrometer to measure spectra from Sirius and some other bright stars, noting the presence of dark lines similar but not identical to those in sunlight. He even succeeded in finding some more bright lines in the light from flames containing particular elements.

The use of diffraction gratings to determine the exact wavelengths of the dark lines gave Fraunhofer what he needed to pursue his interest in "practical optics," namely a set of references by which to better measure the dispersion of various types of glass. He therefore left the field to others, noting only that his experiments had "furnished interesting results in physical optics, and it is therefore greatly hoped that skillful investigators of nature would condescend to give them some attention." He returned his focus to building up industrial glass production until he died young in 1826, like many glassmakers of that era.†

While Fraunhofer's interest in spectroscopy was purely industrial, other scientists did, in fact, follow up on his discovery of spectral lines. In the United Kingdom, William Henry Fox Talbot and John Herschel studied the bright

---

* These days, you can readily buy clear plastic diffraction gratings in the form of novelty "glasses" that produce rainbow patterns when you look through them at small sources of light.
† The official cause was tuberculosis, but it seems likely that this was at least aggravated by inhaling heavy metal vapors from the chemicals used in his glass recipes.

lines in the spectrum of light emitted by various chemical compounds heated in flames and showed that these bright lines could be a useful tool for identifying minute quantities of particular elements. The French physicist Jean Bernard Léon Foucault provided a conceptual explanation of Fraunhofer's dark lines when he demonstrated that a relatively cool vapor of a given element would absorb light at the same wavelengths emitted by that element heated in a flame. The Fraunhofer lines, then, are the result of light from the hot sun being absorbed in the relatively cool outer part of its atmosphere.

The spectral lines of chemical elements began to be studied in a truly rigorous way in the middle 1800s, thanks largely to work by the German physicist Gustav Kirchhoff and chemist Robert Bunsen. They began cataloging the spectra of known chemical elements and showed that each had its own unique pattern of spectral lines emitted when a vapor of that element is heated to sufficiently high temperature. This quickly became a useful tool for identifying new chemical elements, beginning with Bunsen and Kirchhoff announcing the discovery of rubidium and cesium in 1861. By 1868, the practice was sufficiently well established that an entirely new element was identified through a spectrum alone, with no terrestrial sample to study. The discovery was based on a yellow-green line in the spectrum of the sun spotted by the French astronomer Pierre Jules Janssen and independently by English astronomer Norman Lockyer a few months later. This was dubbed "helium" by Lockyer, after the Greek sun god Helios, but helium gas wasn't isolated on the earth until 1895.

The feature that makes spectral lines important for timekeeping, though, is the same one that made them important for Fraunhofer: they are fixed in wavelength, which means they are also fixed in frequency.* The light absorbed or emitted by a particular element has a characteristic frequency that is always the same, and the frequency of the light wave can be used as the basis for a clock. Using the oscillation of light as the "tick" of a clock has the very great advantage of removing material objects from the timekeeping operation: there are no moving parts to run down or wear out.

---

* The frequency of a wave multiplied by its wavelength is always equal to the speed of the wave, regardless of what's waving, and as we've seen previously, the speed of light is a universal constant.

The idea of using spectral lines as fixed-frequency sources to measure time—that is, a clock where the tick is the light wave going from peak to valley back to peak—was first suggested by Lord Kelvin in 1870, but the first atomic clocks wouldn't be built until 1949, more than 120 years after Fraunhofer's death. The reason for this is largely technological—nobody, Lord Kelvin included, knew a good way to reduce the extremely high frequency oscillations of visible light down to something that might drive practical clockwork.* The idea of spectral lines as a time standard also faced a philosophical problem: spectral lines couldn't be a good basis for a clock because nobody understood how they worked. The existence of unique patterns of lines was an empirical fact and undeniably useful for both industrial glassmakers and academic chemists, but their *origin* remained a mystery. Without an understanding of *why* particular atoms absorb and emit particular frequencies of light, it's difficult to have confidence that the frequency is, in fact, constant at the level needed to make a reliable clock.

That understanding didn't begin to emerge until 1913, nearly a full century after Fraunhofer's first observation of dark lines in the spectrum of sunlight. The explanation of spectral lines in terms of the structure of atoms was one of the first and most important steps in the quantum revolution.

The Football Player's Atom

In 1911, Ernest Rutherford did something that surprised and amused his colleagues at the University of Manchester: he hired a theorist. Rutherford was renowned as an experimentalist, having recently won the 1908 Nobel Prize in Chemistry for showing that heavy atoms are transformed into new elements when they undergo radioactive decay and emit alpha particles.† Rutherford was also known to disdain other sciences (the quote "In science there is only physics, all the rest is stamp collecting" is often attributed to him) and the more abstract

---

* It's only in the twenty-first century, with the development of optical frequency combs, that it's really become practical; we'll discuss this more in Chapter 16.
† The irony of Rutherford winning the chemistry Nobel was not lost on him; he joked at the banquet that of all the transformations he studied, none was more surprising than his own from physicist to chemist.

parts of his own field (he's also often pseudo-quoted as saying, "If your experiment needs statistics, you ought to have done a better experiment").*

The theorist in question, a young Dane named Niels Bohr, also seemed a particularly odd match. Rutherford was a voluble New Zealander with a booming voice that sometimes upset the more delicate apparatus used in his laboratories, while Bohr was soft-spoken and equivocal. The two hit it off, though, through a shared interest in sports—Bohr had played soccer at a high level in Denmark (and his brother Harald was on the country's Olympic team in 1908),† and when some colleagues teased him about hiring a theorist, a flustered Rutherford is said to have declared "Bohr is different. He's a *football player*."

As unlikely as their connection may have seemed, Rutherford was badly in need of a theorist, thanks to some experiments done in 1909 by his assistant Hans Geiger and an undergraduate student named Ernest Marsden. The two of them were firing alpha particles at a thin gold foil and were shocked to find that substantial numbers of these bounced off the foil, when the prevailing theory of atoms at the time said this should essentially never happen. Rutherford realized that this result could only be explained if the vast majority of the mass of the atoms were concentrated in a tiny, compact ball at the center, thus creating the now familiar picture of an atom as something like a little solar system. The negatively charged electrons in the atom orbit the positively charged nucleus at a distance of nearly 100,000 times the radius of the nucleus, a difference in scale that makes the nucleus like "a fly in a cathedral."

While Rutherford's model fit the data of Marsden and Geiger, it had one small problem: according to the laws of physics known at that time, it was absolutely impossible. The orbiting electrons in Rutherford's atom are constantly changing their direction of motion as they whip around in their orbits, which means they ought to constantly be emitting electromagnetic waves. Those

---

* As with most great lines in the history of science, it's impossible to find a citation of him actually saying any of these things in print, but they're attributed to him by people who knew him.

† Niels Bohr did not play for the national team and only spent one year on a high-level club, as he was more interested in science. In one of his club games, playing goalkeeper against a German team, he gave up a goal on a long shot that should've been an easy save, which he later admitted happened because he was thinking about a math problem and not paying attention to the action on the field.

waves would carry away energy, causing the electrons to slow down, which in turn would cause them to spiral into the nucleus. In a world governed by classical physics, Rutherford's atomic model is unstable: the orbiting electron would spew out an enormous burst of X-rays before crashing into the nucleus and ceasing to be an atom.

Bohr found a solution to the problems of Rutherford's atomic model by bringing in a new idea from another part of physics, first introduced by Max Planck in 1900. Planck was struggling to explain the spectrum of light emitted by a hot object and found he was only able to make the math work out if he assumed the object contained "oscillators" that could emit light of a given frequency only in discrete amounts of energy. Each oscillator had a characteristic energy equal to the frequency of the light multiplied by a small number (which we now call "Planck's constant" in his honor) and could emit only integer multiples of that energy: one unit, two units, three, units, but never one and a half units or pi units.

This trick got Planck the right formula for the spectrum of thermal radiation, but he always considered it ugly and inelegant and hoped somebody else would find a better way to derive the same formula. Instead, he got the exact opposite: in 1905, Einstein picked up Planck's idea and ran with it. Einstein suggested a "heuristic model" of light with a kind of particle character: a beam of light contains some large number of "light quanta" (these days we call them "photons"), each carrying an amount of energy given by Planck's formula: the frequency of the light multiplied by Planck's constant.

This notion of connecting frequency to the energy of particles of light was the most radical thing Einstein did in his career, but it worked extremely well to explain the photoelectric effect. When Einstein was awarded the Nobel Prize in Physics for 1921, the one specific accomplishment called out was his theory of the photoelectric effect.*

Bohr took the idea of light being emitted in discrete amounts of energy and used it to improve Rutherford's solar-system model of the atom. Bohr suggested that there are certain special orbits (particular combinations of radius and speed

---

* The absence of relativity theory from Einstein's citation surprises a lot of people. It's largely the result of politics within the Royal Swedish Academy of Sciences, who choose the laureates. The photoelectric effect citation was a compromise.

for the electron) in which the electrons simply do not emit light. Each of these orbits has a characteristic energy, and atoms emit or absorb light only when they move *between* these allowed orbits. The frequency of the light emitted or absorbed is determined by Planck's rule: the frequency multiplied by Planck's constant is equal to the difference in energy between the two orbits.

Like the Planck and Einstein models for light, Bohr's atomic model also included an element of discreteness: the special orbits for the electrons inside an atom are defined by their angular momentum being an integer multiple of Planck's constant.* This means each special orbit is assigned a quantum number n, which determines the energy for that orbit, with the energy increasing with increasing n.

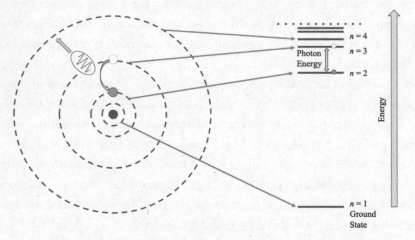

The Bohr model of the quantum atom, with discrete electron orbits whose energy increases with increasing n. When an electron moves between these states (dropping from $n = 3$ to $n = 2$ in the figure), the atom emits or absorbs a photon whose energy is equal to the difference between the energies of the orbits.

Bohr's quantum model of the atom provides a conceptual explanation of why there are spectral lines: for any given atom, there are only a limited number

---

* Angular momentum is a quantity associated with rotational or orbital motion; loosely speaking, it's the momentum (mass times velocity) of the electron multiplied by its distance from the nucleus.

of special orbits with well-defined energies,* so there are a limited number of possible frequencies that can be absorbed or emitted. The model also works very nicely to explain the spectrum of hydrogen (the simplest atom, with only a single electron, whose spectrum had already been shown to fit a pattern where lines were identified with integers) and hydrogen-like ions (atoms from which all but one electron has been removed). It also provides a useful model for the characteristic X-ray radiation emitted by various elements, which was then being studied by another of Rutherford's students, Henry Moseley.[†]

Bohr's model was not well liked by physicists of the older generation, but it worked well enough that a community of younger physicists quickly sprang up, dedicated to working out the properties of the quantum atom, and they had a good deal of success. This period, from Bohr's first papers around 1913 to the development of modern quantum mechanics around 1925, is now known as the "old quantum theory" and represented the first time that physicists could really begin to talk sensibly about spectral lines and the structure of atoms. Bohr returned to Denmark as the first chair of Theoretical Physics at the University of Copenhagen and worked to build up an Institute of Theoretical Physics there.[‡] Bohr's Institute became a central hub for the rapidly developing quantum theory.

Full Quantum Theory

The Bohr model represents a conceptual breakthrough in understanding the physics of atoms and spectral lines (and is sufficient to explain the basic

---

* Technically, there are an infinite number of possible special orbits, one for every integer, but the energy increases toward a maximum value, crowding closer and closer together at higher values of n. In practice, this means that only a smallish number of the lower-energy states are separated by energy differences that correspond to frequencies associated with visible light.

† These X-rays are created when a sample is bombarded with high-energy particles, which can knock loose some electrons from the innermost orbits. When these are replaced by electrons that started in high-energy states, they emit X-ray photons. Moseley found that there was a characteristic maximum energy for each element that followed a simple pattern, which can be explained in terms of the lowest-energy Bohr orbit.

‡ The Institute was funded in part by the Carlsberg beer company, and formally opened in 1921, just before Bohr won the 1922 Nobel Prize in Physics. The Carlsberg connection led to a nice perk for Bohr: after he won the Nobel, the company gave him a house and an unlimited supply of beer piped in from their brewery next door.

principle behind atomic clocks), but there are a number of problems with the "old quantum theory." In particular, it's not entirely obvious why the particular states singled out by Bohr should be special. A deeper theory was needed, and in the mid-1920s, one was found but turned out to be far stranger than anyone anticipated.

One path to understanding the development of modern quantum mechanics runs through France, where a young PhD student from an aristocratic family made an intriguing suggestion in 1924. Louis Victor Pierre Raymond de Broglie (normally just "Louis de Broglie") was taken by Einstein's idea of light having particle character, and in his doctoral thesis submitted to the University of Paris suggested a sort of symmetry: if light, which everyone knew was a wave, has particle character, then perhaps an electron, which everyone knew was a particle, should have wave character. He suggested that there is a wave associated with the electron, having a wavelength that's inversely proportional to its momentum (i.e., if you double the momentum, you cut the wavelength in half). This closely parallels the relationship between wavelength and momentum in light, making both light and matter follow similar rules.

The idea of wave nature for electrons is a bizarre one, and initially his professors had no idea what to make of it. One popular story says they dithered about whether to award de Broglie his degree until after they showed his thesis to Einstein, who pronounced it brilliant. The wave nature of electrons was demonstrated directly in the late 1920s, in experiments that showed electrons interfering with each other in the same way that light waves do. Well before that, though, the idea had strong conceptual appeal, as it offers a way to understand what makes the Bohr orbits special.

If you put de Broglie's electron wavelength together with Bohr's requirement of integer angular momentum, you find an intriguing relationship: the circumference of the orbit is an integer multiple of the wavelength of the electron in that orbit. This means that when the electron wave comes back to where it started, it's at the same phase of its oscillation: the peaks of the next orbit fall in the same places as the peaks of the previous orbit. This sets up a situation where the orbiting electron can constructively interfere *with itself*, making what's called a "standing wave." This is similar to the physics that picks out the characteristic frequencies found in musical instruments, another situation in which infinite possibilities get trimmed down to a countable set of possibilities.

This similarity to a familiar problem makes it an appealing way to think about the origin of the special orbits in Bohr's atom: Bohr's quantum number n is then just the number of peaks in the electron standing wave.

There are significant problems with de Broglie's original approach to the wave nature of electrons, but his suggestion set other physicists on the path toward a more correct theory. In particular, if the electron has wave nature, that suggests there ought to be a wave equation that governs its behavior; this idea set the Austrian physicist Erwin Schrödinger looking for such an equation, which he found while on a ski holiday at the end of 1925. Right around the same time, one of Bohr's proteges, the German physicist Werner Heisenberg, found a more abstract approach to the problem of atomic structure that produced the same answers.* While Schrödinger's wave equation and Heisenberg's "matrix mechanics" use very different mathematical tools, they turn out to be perfectly equivalent, and physicists nowadays readily switch back and forth between them, using whichever is most convenient for a particular problem.

There are numerous books going into great detail about the predictions of quantum mechanics,† which are weird and fascinating, with consequences that are still the subject of active research more than a hundred years after the appearance of the first quantum models. For the purposes of our discussion of atomic clocks, though, we only need to know about a few of the most important high-level ideas. All of these ultimately flow from the idea introduced by de Broglie and formalized by Schrödinger, that material objects have both particle and wave character.

The most important of these elements is the idea that every quantum object is described by a wavefunction, a mathematical object that changes in time according to the Schrödinger equation. As the name suggests, this has wavelike properties (it extends over a region of space and oscillates in time at a frequency that depends on the energy of the object it's describing), but it's not directly

---

* Like Schrödinger, Heisenberg's great inspiration came while he was away from home, in his case, escaping severe allergies by spending time on the remote island of Heligoland. The examples of Schrödinger and Heisenberg have been used ever since by physicists arguing for more vacation time.

† See, for example, *Breakfast with Einstein: The Exotic Physics of Everyday Objects* (BenBella Books, 2018).

observable itself (we can't look at a single electron and see the waviness associated with the wavefunction).

What the wavefunction ultimately tells us is the *probability* of finding a quantum particle in a particular state: that is, in some particular position, with some particular energy, etc., depending on what's being measured. We get the probability by an operation that's not too far different from squaring the wavefunction, and this probability distribution is ultimately the only thing we can predict. What's "really" going on in the process that takes us from the wavefunction to the result of a single measurement remains an open question, and a highly contentious one; there are many different interpretations of quantum mechanics, each passionately supported by some group of physicists. Numerous books have been written about these interpretations, but this is not one of them: for our purposes, all we need to know is that quantum objects are described in terms of probability.

Thanks to this probabilistic nature, quantum objects can exist in multiple states at the same time, until they are measured. Such "superposition states" are, again, a feature that comes from the wave nature of everything: in the same way that a bit of air in a concert venue can be vibrating at multiple frequencies at once due to the different notes played by different instruments, a quantum atom can be in multiple energy states simultaneously. The total wavefunction is the sum of all the different waves that contribute to the total state. As with waves in air, the resulting function can be very complicated and show interference effects from all those different contributions. This quantum interference is critically important for the operation of state-of-the-art atomic clocks.

The probabilistic nature of quantum mechanics does not, however, mean that its predictions are totally up for grabs. The wavefunctions and probabilities that quantum physicists calculate are highly accurate and tested against ultra-precise measurements of real physical quantities. For one particular quantity having to do with the magnetic behavior of electrons, a number called the "*g*-factor" for the electron, the theoretical prediction and experimental measurement agree perfectly to 11 digits.* This phenomenal accuracy gives physicists a

---

* The predicted value is 2.002319304363286 ± 0.00000000001528, and the best current measurement is 2.00231930436146 ± 0.00000000000056. The measurement is actually slightly more precise than the prediction, which reflects the extraordinarily complex calculation required. The analogous quantity for a "muon," an exotic particle similar to an

high level of confidence in the stability of the quantum states in atoms, allowing them to serve as the basis for the most accurate clocks in history.

FROM PENDULUMS TO ATOMS

From the late 1600s all the way into the twentieth century, the most accurate timekeepers available were all pendulum clocks. Generations of clockmakers tweaked the basic design little by little, refining the escapements and materials and making other adjustments to remove tiny imperfections that would change the rate of ticking.

The pinnacle of the mechanical clockmaking art was reached at about the same time quantum physics arrived on the international scene. The Shortt-Synchronome free-pendulum clock was developed in 1921 by William Hamilton Shortt and Frank Hope-Jones, and actually consists of *two* pendulums. The "master pendulum" is made of an alloy with exceptionally low thermal expansion and swings inside a sealed, temperature-controlled chamber from which all the air has been removed. This is not directly connected to any clockwork; instead, a small electric sensor is used to check its position every 30 seconds and compare it to a secondary pendulum swinging in air, which *is* connected to the clockwork. This two-pendulum scheme minimizes perturbations to the oscillation of the master pendulum while keeping the secondary pendulum in synch with it; the secondary pendulum is connected to the gears and other elements needed to provide a clear time readout. The combination produced a clock good to better than one second per year; a modern reconstruction using updated sensors was good to about one second in 12 years.

The widespread use of Shortt pendulum clocks at national standards laboratories and observatories eventually made clear a problem that had been suggested by astronomical observations: the rotation of the earth is simply not stable enough to serve as a basis for time measurement. It had been known for millennia that the length of a solar day varied over the course of a year,

---

electron but much heavier, disagrees with theory by a small amount (as of early 2021), and is regarded by some as a possible hint of "new physics" beyond the particles and interactions that we currently understand.

a problem that was addressed by using a "mean solar day" that averaged the lengths over a full year, and then converting between the apparent solar time and this mean time using the "equation of time" to account for the variation on any particular date.

In parallel with the steady improvement of pendulum clocks, astronomers had continued to refine both measurements and predictions of the positions of the planets in our solar system, producing ever better versions of the tables needed for celestial navigation. The high point of this development was the work of the American astronomer Simon Newcomb, who in 1895 compiled decades of measurements (his own and those of other astronomers) to make his *Tables of the Motion of the Earth on Its Axis and Around the Sun* (more commonly referred to as "Newcomb's *Tables of the Sun*"), which predicted the position of the earth in its orbit with unprecedented accuracy.

Newcomb's tables gave the best measurement to that point of the mean solar day and were used to produce navigational tables giving the predicted positions of the sun and planets. There was some hint even in this work, though, that the length of the day might not be perfectly stable: that is, that the *mean* day also changed length over time. With the development of Shortt clocks,* the evidence became undeniable.

The shift in the length of a mean day is very small—a day in 2019 was around two milliseconds longer than in 1870—but over very long spans of time can be significant. Ancient records of eclipses of the sun and moon allow modern astronomers to determine the slow lengthening of the day by comparing the observations to predictions that "run the clock back" using the modern length of a day. For example, a total eclipse recorded in Babylon in 720 BCE would've been visible only over the Atlantic Ocean if the length of a day then were the same as now. The fact that this eclipse was visible in modern-day Iraq tells us that the mean day has gotten longer by a bit less than two milliseconds per century; all those milliseconds over all those days add up to several thousand miles of the earth's rotation.

This long-term slowing is mostly due to the moon. As the moon's gravity pulls on the earth, it causes small distortions of the planet and a large-scale

---

* And also quartz crystal clocks, at around the same time, which we'll discuss briefly in Chapter 15.

shifting of the oceans that we see as the pattern of daily tides. These tidal effects act to resist the rotation of the earth, gradually slowing it, and in response the moon shifts slightly farther away. Laser ranging experiments using reflectors left on the surface of the moon by the Apollo missions confirm that the moon is, in fact, slowly moving away at a rate of around 3.8 centimeters per year.

In addition to the gradual slowing caused by tidal forces, the earth's rotation rate is subject to more sudden changes, caused by the shifting of continental plates and the motion of material in the earth's mantle. This can sometimes be traced to single dramatic events—for example, the magnitude 9.0 earthquake off the coast of Japan in 2011 is believed to have shortened the length of a mean day by about 1.8 microseconds from its pre-quake value. More often, though, the changes are subtle and slightly mysterious, detected after the fact by ultra-precise astronomical measurements.

A clock good to a second per year can detect changes of a few milliseconds in the length of a day, so with the development of the Shortt pendulum clocks, human-made clocks reached a level of reliability in timekeeping that exceeded that of the earth's rotation. By the 1940s, scientists recognized a need to do better than defining a second as 1/86400th of the time needed for a mean day.

The first shift away from a rotation-based day was made using Newcomb's *Tables of the Sun*, since the earth's orbital motion is much more stable and reliable than its rotation. Accordingly, in 1960, the definition of a second was officially changed to "the fraction 1/31,556,925.9747 of the tropical year for 1900 January 0 at 12 hours ephemeris time." This "ephemeris time" was based on the position of the sun relative to the stars as calculated from Newcomb's tables, moving the definition of time away from the irregular motion of real objects to a more ideal solar system rooted in Newtonian physics.

The many digits in the denominator of that fraction mostly reflect the high degree of precision with which the tropical year is known, thanks to millennia of observation. It is worth noting, though, that the value used to calculate the ephemeris second also differs very slightly from the number of seconds one would expect using days of exactly 86,400 seconds and the length of the tropical year from Newcomb's tables. The difference is very small—the digits after the decimal point are "9747" instead of "9754"—but reflect the need to mesh with the mean-solar-day second that had been used to that point.

The era of ephemeris times was only a temporary fix, though, because the quantum connection between energy and frequency offers the possibility to define time in a way anchored in the fundamental laws of physics—and track it with a clock that contains no moving parts.

HOW AN ATOMIC CLOCK WORKS

In a sense, the name "atomic clock" is a bit misleading; these devices might more accurately be called "light clocks," since the oscillating element whose "ticks" are counted to mark time is microwave radiation. The atoms then act as a reference to ensure the microwaves are oscillating at the correct frequency. The process of operating an atomic clock recapitulates the basic procedure for checking the performance of any clock: the microwaves are synchronized with the atoms, both atoms and microwaves are allowed to operate freely for some time, and then they are checked against one another to see if they're still in synch. If they're not, the microwave frequency is adjusted and the process repeated.

The idea of using microwaves for time measurements was first publicly proposed in 1945 by the American physicist Isidor I. Rabi, who based it on the molecular beam spectroscopy technique that won him the Nobel Prize the year before. The first "atomic" clock was actually based on a molecule, with a system built at the US National Bureau of Standards in 1949, using a transition between two states in ammonia to demonstrate the soundness of the general principle. The first truly atomic clock, using cesium, was built by Louis Essen and Jack Parry of the UK's National Physical Laboratory in 1955. Just over a decade later, cesium was officially enshrined as *the* atomic standard, with the second defined as "the duration of 9,192,631,770 periods of the radiation corresponding to the transition between the two hyperfine levels of the ground state of the cesium-133 atom." As with the change from mean solar time to ephemeris time, the many digits given in the definition were chosen to closely match the ephemeris-time second in use at the time of the switch.

The cesium frequency used to define the second is very high, but that's a good thing for precision measurement. Counting "ticks" to measure a time interval is inherently a discrete process—you're not going to do all that much better than plus or minus one tick—and because of that, the more ticks you can count, the better your final measurement will be. This is the principle that allowed the makers of the Gregorian calendar to determine the length of the year to within a few minutes based on observations of the solstices and equinoxes that were only good to about a day: they were drawing on thousands of years of recorded data. The cesium clock brings this same principle down to the level of the second. If you can make a good count of the number of oscillations of the light in question—something it was possible to do thanks to microwave technology developed through the World War II radar program—you can determine the length of a second to exceptionally high precision. Measuring one second by counting cesium oscillations with an accuracy of plus or minus one gets you the same measurement precision you would get by determining the length of a tropical year from counting the days in 25 million years.*

Atomic clocks have gone through a lot of technical advances in the last half century, though cesium remains the standard. State-of-the art atomic clocks these days operate in a "fountain" configuration, beginning with several million cesium atoms cooled to within a few millionths of a degree of absolute zero.[†] The atoms are prepared in one of the two states specified in the SI definition of the second, then launched upward at a speed of a few meters per second. A short distance above the launch point, they enter a microwave cavity, a hollow metal chamber whose dimensions are chosen to support a standing wave at the microwave frequency associated with the cesium transition where they interact with microwaves for the first time.

The parameters for this first interaction—the size of the cavity, the launch speed of the atoms, and the strength of the microwaves—are chosen so that the microwaves move the atoms halfway from one state to the other. This doesn't mean simply moving half of the atoms between states, though, leaving the other half behind. Instead, each individual atom is prepared in a quantum superposition of both states at the same time.

This may seem like a strange thing to do, but it dramatically increases the sensitivity to small frequency differences, thanks to the physics of interference.

---

* Of course, as we've just seen, this would be futile given the changes in the length of the mean solar day, but it gives a sense of the scale.
[†] There's a lot of fascinating physics involved in getting to this point, but explaining it all could easily take another whole book.

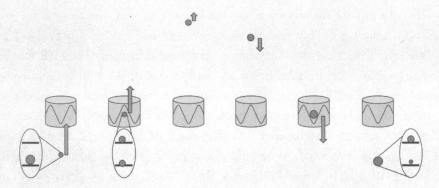

Fountain clock process. A ball of cold atoms is prepared in the lower of the two energy states, then launched upward with some speed. They pass through a microwave where they're prepared in a superposition of both states, then fly up above the cavity where they slow, stop, and fall back down. After the second pass through the microwaves, the state is measured again to determine the probability the atoms made a transition between states.

We can think of the microwave interaction as being like the beam splitter in a Michelson interferometer,* taking the single incoming wavefunction associated with the atom and splitting it into two pieces, one corresponding to each state. These will eventually be recombined (spoiler!), and what happens at that point will depend on the relative phase between the two pieces.

Unlike the light in a Michelson interferometer, though, where the waves on each arm oscillate at the same frequency, the two pieces of the wavefunction corresponding to the two different states evolve at different rates. As a result, even though they're not physically separated, the phase *difference* between the two states changes at a rate that exactly matches the microwave frequency associated with their energy difference. This first interaction, then, is essentially creating a tiny clock *within* the atom, one that starts out synchronized with the microwave source.

After the atoms pass through the cavity, they continue to rise, slowing under the influence of gravity, reaching a maximum height of a meter or so above the cavity before turning around and falling back down. During this time, the internal "clock" of the superposition state is changing phase, and so is the "lab clock" of the microwave source. About a second after the upward pass

---

* As discussed in Chapter 12.

through the cavity, the falling atoms reenter the cavity on their way down and interact with the microwaves for a second time.

This second pass through the cavity functions like the second encounter with the beam splitter in a Michelson interferometer, bringing the two pieces of the wavefunction back together. What happens in that recombination will depend on the phase difference between the two pieces of the wavefunction, and the frequency of the microwave source. In this case, though, the different outputs are not different physical directions but the two possible states for the atoms.

If the frequency of the microwaves exactly matches the energy difference between the two states in the atom, the internal clock created by the superposition and the lab clock will still be perfectly in synch. In this case, we see a kind of constructive interference between the wavefunctions for the two states, and the second pass through the cavity completes the process started by the first: all of the atoms will move to the other state.

If the frequency of the lab clock is a little bit off, though, the phase difference between the atoms and the microwaves increases, and the interference is no longer perfectly constructive, so the number of atoms changing states decreases. As the difference increases, at some point the interference between states becomes perfectly destructive, and *none* of the atoms make the transition from one state to the other. If you continue to increase the frequency difference, eventually the two are out of synch by one full oscillation, and you get constructive interference again, then destructive, and so on.

As noted above, quantum physics can predict only probabilities, but each of the millions of atoms in the fountain goes through the same process independently. A single measurement of how many atoms are in each of the two states after the full fountain cycle thus tells you that probability to high accuracy. A graph of the fraction of the atoms that change states as a function of frequency, then, will show an oscillation between 0 and 100 percent as the interference between states goes from constructive to destructive and back. These oscillations are called "Ramsey fringes" after Norman Ramsey (a PhD student of Rabi's before World War II), who devised this two-step method for atomic clocks in 1949, for which he received a share of the 1989 Nobel Prize in Physics.*

* Ramsey got half the prize, and the other half was split between Hans Dehmelt and Wolfgang Paul for the development of techniques for trapping ions.

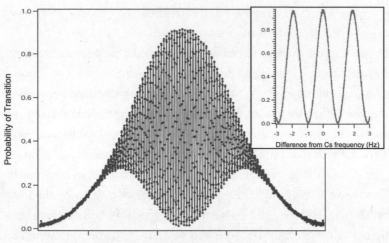

"Ramsey fringes" from an atomic clock at the US Naval Observatory, showing
the rapid oscillation in the probability of moving between states for different
microwave frequencies. As depicted in the inset, the probability goes from
near 100 percent to near zero with a frequency change of just 1 Hz.

The width of these fringes—that is, the change in frequency needed to go
from 100 percent excited to 0 percent—decreases as the time between interactions increases. This is the primary feature that makes the fountain geometry
attractive: it allows for a very long time between interactions, and that increases
the sensitivity to small changes in frequency.* For a modern cesium fountain
clock, a single relatively crude measurement is good to better than +/– 1 Hz out
of the cesium frequency of 9,192,631,770 Hz; more sophisticated techniques
can significantly improve on that precision even for a single clock cycle. A continuously operating clock at that level of accuracy would gain or lose no more
than 0.003 seconds in a year.

But, of course, a real atomic clock isn't used to make just a single measurement. True clocks are operated nearly continuously, making repeated measurements and small adjustments to the frequency. These small tweaks combine to

---

* A secondary, not-insignificant benefit of the fountain technique is that it uses the same
  microwave cavity twice, removing the need to make two identical cavities, which was a
  source of uncertainty in older atomic clock systems that used a fast-moving beam of atoms.

TIME FOR THE WORLD

The preceding description explains the operation of a single state-of-the-art atomic clock, which in principle could be sufficient to produce a time signal accurate enough for almost any human purpose. But, of course, operating only a single clock for the whole world would be wildly impractical for any number of reasons: a single-clock time would be vulnerable to breakdowns or power outages, and it would potentially create conflicts over issues of national pride. (Where is the one world clock to be housed?) Most importantly, though, an average of many measurements can be more precise than any of the individual measurements in the average. For all these reasons, the actual official time for the world is provided by a global ensemble of several hundred atomic clocks operated by more than 80 national laboratories.

Any operation with that many contributors will necessarily involve a certain amount of bureaucracy, which brings with it a collection of confusing jargon and acronyms. In this case, the process of determining time for the world is overseen by the Bureau International des Poids et Mesures (French for "International Bureau of Weights and Measures," usually abbreviated BIPM), which produces a number of time standards for reference, of which only two matter for our purposes: the International Atomic Time (or TAI from the French form Temps Atomique International) and Coordinated Universal Time (UTC*). This process is necessarily a retrospective one: once a month, the Time Department of BIPM sends out a document known as "Circular T," which tells you what time it *was* up to a month earlier.

The way the process works is yet another recapitulation of the synchronize–free-run–check cycle that's so fundamental to timekeeping. Each of the contributing laboratories maintains its own array of clocks—some atomic fountains of the type described above, some older atomic clocks using a beam of atoms passing through two different cavities—and combines them locally to generate an approximation of UTC, designated with a code in parentheses to indicate the lab from which it originates: thus, the US Naval Observatory

---

* This functions as a sort of diplomatic compromise between the English CUT and the French TUC (from Temps Universel Coordonné), but it's also consistent with the previous convention of designating variants of "Universal Time" based on the earth's rotation. For example, the mean solar time at 0 degrees longitude is designated UT1.

produces UTC(USNO), while the German Physikalisch-Technische Bundesan-stalt produces UTC(PTB). These are distributed directly and continuously to clients who need a time reference signal delivered "live," as it were—for exam-ple, UTC(USNO) is distributed to the US military for their official purposes and also provides the time signal for the Global Positioning System (GPS) satellite network (more about this just ahead), while the National Institute of Standards and Technology provides UTC(NIST) for civilian uses, including a handy Internet Time Service you can use to synchronize your computer's clock with official atomic time.

These "live" realizations of UTC are more than sufficient for most every-day purposes, but certain high-precision applications, mostly in astronomy and geodesy, need a time reference that's more consistent than the individual real-izations. This is where the parentheses-less UTC comes in.

At regular intervals, each lab records the difference between each of their clocks and their local realization of UTC, and also the difference (in nanosec-onds) between their local realization of UTC and some other version; for con-venience, this mostly uses the GPS network. All of these differences—between local clocks and local UTC and between different versions of local UTC—are communicated to BIPM, which can then use basic arithmetic to compute the difference between any two clocks in the whole international collection.

Using these data, BIPM does two things: it evaluates the performance of each clock, and it calculates an average time. These two processes are interre-lated, as better-performing clocks are given greater weight in the final aver-age. As there's no "correct" time to be targeting, the evaluation of clocks is more about stability than the absolute difference: a clock that's consistently 100 nanoseconds behind will be given greater weight than one that's some-times five nanoseconds ahead and sometimes five nanoseconds behind. Once the averaging process is completed, the resulting UTC timescale is published in Circular T as a list of differences between the final average UTC and the local UTC for each of the 80-odd laboratories contributing times to BIPM. Users who require the extra precision of the full UTC ensemble will record their time measurements based on one of the local realizations, and then retrospectively correct their data to the official UTC time.

The final element of this process is the one we opened this chapter with: the leap second. As we know, the rotation rate of the earth is not constant, and as a result, time as tracked by atomic clocks does not perfectly match time of day in an astronomical sense. The atomic definition of the SI second was chosen in 1967 to closely match the previous definition, which in turn was based on the ephemeris prediction for the year 1900. As a result, the atomic-clock second is not a great match for the more conventional sense of a "second" as 1/86,400th of a modern solar day.

Left alone, this difference would slowly accumulate, in the same way that differences between the tropical year and the Julian calendar did, eventually leading atomic-clock-based time to drift out of synch with our experience of the day—the sun would rise just after midnight and set just after noon. To avoid this problem, another agency, the International Earth Rotation and Reference Systems Service (IERS),* monitors the motion of astronomical objects to determine the difference between atomic-clock time and rotating-Earth time. When that difference gets to 0.9 seconds, IERS will inform BIPM, and a leap second will be added to UTC on either June 30 or December 31.[†] This happens at irregular intervals; 27 leap seconds have been added since the current system was implemented in 1972, the most recent (as of August 2020 when I'm writing this) in December 2016. For the purposes of observations where the difference between measurements is important but the exact orientation of the earth is not, there is the TAI timescale, which is just UTC without the leap seconds.[‡]

As complicated as this may all seem, this is if anything a streamlined description of the process. There's also a process of "steering," whereby some labs will attempt to adjust their local UTC realization to more closely match the final average. There's even some political controversy around the whole idea of leap seconds, because they present some technical challenges for computer networks and the like. Proposals have been floated to do away with leap seconds and either let UTC drift relative to the rotation of the earth or plan to

---

* Formerly simply the International Earth Rotation Service, which sounds like the people you employ to periodically wind up the big spring that keeps the earth spinning.

† In principle, a leap second could be added at the end of March or September as well, but it's never been needed.

‡ The difference between UTC and TAI is actually 37 seconds, which includes a 10-second difference that had accumulated between the start of atomic time and the implementation of the leap-second system.

make larger adjustments at much longer intervals. There are even a couple of non-BIPM timescales, mostly used in astronomy, that correct for relativistic effects of the earth's motion,* which adds another level of complexity.

At the moment, though, UTC represents the most accurate and reliable timekeeping system that's been developed in the last several thousand years of humans tracking the passage of time. Our ability to measure time using atomic clocks has developed to the point where all of our other measurement standards are now based on time measurements, giving fixed values to fundamental constants. So, for example, the speed of light is defined as 299,792,458 meters per second *exactly*, with the meter then defined as the distance light travels in 1/299792458th of a second. Other quantities are similarly defined by giving fixed values to fundamental quantities and associating them with time measurements: the charge of an electron is given a fixed value, and current is defined as some number of those charges per second, etc. The last physical artifact standard, a platinum-iridium cylinder that had defined one kilogram of mass since 1889, was replaced in 2019 by fixing the value of Planck's constant connecting energy and frequency.[†] This definition allows the kilogram to be found from Planck's constant, the speed of light, and the second as determined by an atomic clock.

---

* These are Geocentric Coordinate Time (TCG) and Barycentric Coordinate Time (TCB), which record time according to an imaginary observer at the nonrotating center of the earth or the center of the solar system, respectively.

† Energy and mass are, of course, related by the World's Most Famous Equation, $E = mc^2$.

produce a microwave clock in the lab that's far better on average than any individual measurement. The fountain clocks that contribute to the official atomic time are typically good to around one part in 10^{16}; that is, one second reported by such a clock is 1.0000000000000000 +/– 0.0000000000000001s. At that level of accuracy, a clock would need to be run continuously for several hundred million years before gaining or losing a single second!

A Sense of Where You Are: Clocks and GPS

Someone confronted with this history of redefinitions and the complexity of UTC might well ask, "What's the point?" After all, for most everyday purposes it's sufficient to define a second as the time needed to say, "One Mississippi," so why bother with cesium fountains and all the rest?

While most of the uses of UTC are admittedly esoteric, there is one application of atomic time that has become an indispensable part of modern life, and it's the same application that has driven precision timekeeping for centuries: navigation. The Global Positioning System (GPS), used by smartphones and navigation systems to provide location tracking in real time, is based on a network of at least 24 satellites carrying atomic clocks and broadcasting the current time. These enable a handheld receiver to calculate the user's current position on the earth to within a few meters.

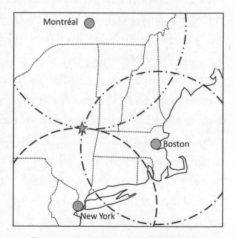

To understand the process, it's useful to consider a slower analogue, in one of the ways I specify the location of Schenectady, New York (where Union College is located), to people who aren't familiar with the area. As a start, I can say that Schenectady is a three-hour drive from New York City, which places it somewhere on a rough circle with a radius of 180 miles centered on Manhattan. Of course, that

The process of locating Schenectady, New York (star), by reference to travel times from better-known cities. The circles represent points three hours from Boston and New York, and four hours from Montréal; Schenectady is at the point where all three intersect.

circle includes a lot of points in Massachusetts, Pennsylvania, and Rhode Island, so I can narrow it further by saying that Schenectady is *also* three hours by car from Boston. Those two travel times put us at an intersection between the circle centered on Manhattan and a similar circle drawn around Boston. There are two such points, one of which is Schenectady, while the other is in the North Atlantic. Common sense probably suffices to distinguish these, but if necessary we can add in the information that we're also a four-hour drive from Montréal; only one city on Earth sits at the intersection of all three of those circles.*

The operation of GPS is similar in specifying distances by time. The orbits of the satellites in the GPS constellation are set up so that at any given time at least three to four satellites are visible from any point on the surface of the earth. Each satellite broadcasts a coded signal identifying itself and giving the time according to its onboard atomic clock. A receiver on the ground uses the differences in the time signals it picks up to determine the travel time for the radio waves coming from that satellite; since the speed of light is constant, this gives the distance to that satellite. This process effectively defines a sphere around each of the satellites, and all of the spheres will intersect at only a single point on the surface of the earth, thus specifying the location of the receiver.

Given the high speed of light (very nearly one American foot per nanosecond, about the only place that measurement system is superior), specifying position to an accuracy of a few meters requires the time broadcast by the satellites to be good to within several nanoseconds, which is why they use onboard atomic clocks. The satellite clocks are regularly updated to follow UTC(USNO), the local realization of UTC generated by the US Naval Observatory. This incidentally makes them a useful part of the process of assembling UTC: any laboratory around the world can readily pick up the GPS signal and determine UTC(USNO) from it, greatly simplifying the comparisons between clocks that are reported to BIPM.

The GPS constellation is also one of the great practical proofs of the theory of relativity. The orbiting clocks are moving very rapidly compared to clocks on the ground, which would make them run slow by around 7 microseconds per day (due to the effects of special relativity discussed in Chapter 12). They are

---

* This is, you may notice, essentially the same procedure used to locate the positions of planets using angular separations from reference stars that we discussed back in Chapter 7.

also at high altitude, which leads them to run *fast* by around 45 microseconds per day (due to the effects of general relativity, which we'll explain in Chapter 14). The engineers designing the GPS satellites took these effects into account and built in a correction factor—a GPS satellite clock operated at ground level would run slow by about 38 microseconds per day. The impressive timing accuracy of the final system thus serves as confirmation of the predicted effects of motion and gravity on time itself.

As good as they are, though, cesium clocks are not the final word in quantum timekeeping. Physicists in labs around the world are working on a variety of schemes to do *even better* than the time provided by modern cesium clocks, using different atoms and techniques, and we'll examine why and how in Chapter 16. First, though, we will check back in on Albert Einstein and explore the happiest thought of his life.

Chapter 14

TIME AND GRAVITY

On November 10, 1919, the front page of the *New York Times* reported scientific results from a British expedition to the island of Príncipe off the coast of Africa to observe a total solar eclipse in May of that year. The story is topped with some of the greatest headlines of all time:

LIGHTS ALL ASKEW IN THE HEAVENS
Men of Science More or Less Agog Over Results of Eclipse Observations

and

EINSTEIN THEORY TRIUMPHS
Stars Not Where They Seemed or Were Calculated to Be,
But Nobody Need Worry

This is the moment that truly catapulted Einstein to the global celebrity status he was to have for the rest of his life, and long beyond it. The theory he published in 1905, now known as "special relativity," was closely paralleled by the work of others; scientists recognized its importance, and it helped Einstein get a series of good academic jobs. If he had stopped there, he likely always would've had to share billing with Lorentz and maybe Poincaré. But of course, he didn't stop there . . .

The 1919 eclipse observation was a dramatic confirmation of an expanded theory that was Einstein's alone, which we now know as general relativity. Completed in 1915, the theory of general relativity built an elaborate mathematical apparatus on the central idea of his 1905 theory: that the laws of physics do not depend on the motion of the observer. Special relativity introduced the idea of time as an individual experience, determined by the speed relative to the observer. The laws of physics are the same for all observers, but the timing of events and the size of objects depends on how fast you're moving.

General relativity says that your experience of space and time depends on *where you are*, specifically how close you are to a massive object. This is an idea with profound philosophical consequences, but one that also has practical implications for people who make and use modern ultra-precise clocks. General relativity is not just a theory of motion, it completes the redefinition of space and time, and in the process provides the key to understanding the operation of gravity. In particular, it predicts that gravity should warp space and time, bending the path followed by light, which is what caused the apparent shift in the positions of the stars near the sun during the eclipse that the *New York Times* reported so breathlessly.

EINSTEIN'S DAYDREAM

As Einstein was wrapping up his formal education at the Swiss Federal Institute of Technology (ETH Zürich), he did what many young scholars do and attempted to secure a teaching position. This is never particularly easy, and faced with both strained relationships with his former professors—Einstein was not exactly subtle about his belief that he was smarter than they were—and the virulent anti-Semitism common in Europe at that time, Einstein was unable to find employment in academia. With the help of a classmate, he was finally able to secure his job as a patent examiner in Bern, starting in 1902.

The work of a patent clerk was fairly routine, so Einstein could power through his assigned duties in short order, leaving him ample time for thinking about physics. He also spent many evenings with a close group of other academically inclined friends—they jokingly dubbed themselves the "Olympia Academy"—reading and discussing the latest developments in science and philosophy. Einstein always credited this environment with helping him to his "miracle year" of 1905, when he published four physics papers, any one of which could have made his career.*

These papers did eventually secure him a series of professorships, ultimately landing him at one of the most prestigious universities in Europe, but nothing in academia happens quickly, so he remained at the patent office for seven long years, reviewing applications and daydreaming. It was here, sometime around 1907, that he had what he later called "*der glücklichste Gedanke meines Lebens*," the happiest thought of his life, while thinking about a person falling off a building.

What separated Einstein from a typical worker bored by a dreary office job is that he was not imagining a *specific* person falling (or being pushed) out the window, but thinking about the physics of the general scenario. In particular, he was struck by the realization that a falling person feels weightless, which led him to connect gravity to relativity.

Einstein's 1905 paper, like the work of Lorentz and Poincaré before him, concerned the effects of motion at constant relative velocity. This is why the theory is called "special relativity," because it deals with the special case in which neither the speed nor the direction of motion is changing. The core of the theory is sometimes stated as a prohibition against distinguishing between states of constant-velocity motion: there is no experiment you can do that can distinguish whether you're truly at rest or moving in some unchanging direction at constant speed.

Of course, we live in a universe full of objects that are constantly changing their speed or direction of motion, but extending the argument of relativity to encompass such accelerating motion is not straightforward.† For one thing, accelerating motion is pretty clearly distinguishable from constant-speed

---

* One of these was the previously discussed paper on special relativity, and another expanded on that to introduce the equivalence of mass and energy. A third was on Brownian motion and helped settle once and for all the question of whether atoms were real, physical entities or merely a convenient calculation tool. The final paper was his "heuristic model" of the photoelectric effect, which helped launch quantum mechanics (discussed in Chapter 13).

† An important note on terminology: while in common parlance, "accelerate" generally means "increase speed," in physics it is used more broadly, to include both speeding up and slowing down, and also changes of direction.

motion, as anyone who's ever ridden in a car can tell you. If you close your eyes, there's no discernable difference between driving down a straight, smooth road at constant speed and sitting still with the engine running, but you can definitely tell when you're accelerating. When the car speeds up, you feel pressed into the seat; when it slows down, you feel pushed forward into the seat belt; and when you turn a corner, you feel pushed toward the outside. These "noninertial forces" make a clear distinction between the apparent laws of physics when you're accelerating versus the laws when you're not changing speed or direction. This may make it seem like the principle of relativity can't be extended to accelerating motion.

Einstein's happy realization at the patent office was that the principle of relativity *can* expand to include acceleration, if at the same time you extend the physics you're considering to include the force of gravity. This extension is now known as the "equivalence principle,"* and it's often expressed in a way that parallels the formulation of the principle of relativity above: there is no experiment you can do that can distinguish whether you're experiencing a gravitational force or just accelerating motion.

The falling person in Einstein's daydream is one example of the equivalence principle in action: they feel weightless because the sensation we experience as "weight" is due to the action of gravity pulling down on our bodies and the ground pressing up on our feet. In free fall, as we accelerate downward at 9.8 m/s^2, those sensations are absent, so we feel weightless, as if gravity has ceased to act. The same thing works in the other direction: the sensation of acceleration you experience when you step on the gas at a stoplight is exactly the same as if gravity had suddenly strengthened and changed direction to pull from the back of the car.

Starting from the equivalence principle and rigorously working out its consequences let Einstein show a number of remarkable results, including the

---

* This terminology is because as a formal mathematical statement it says that the "mass" used to calculate the force of gravity on an object and the "mass" used to calculate the acceleration of an object due to an applied force are identical. This had been widely assumed to be true since Galileo's day for reasons of mathematical convenience, and, like the constant speed of light, seemed to be an empirical fact based on a century or so of high-precision experiments. Einstein elevated equivalence from an empirical observation that happens to be true to the status of a fundamental principle that *has* to be true.

prediction that light should bend under the influence of gravity. This is what was tested by the eclipse expedition that got the *New York Times* so worked up. The same logic also leads to a prediction more directly relevant to the topic of this book: that clocks at different altitudes tick at different rates.

MAGIC ELEVATORS

To understand how gravity affects time, it's useful to start with some simple examples of the equivalence principle in action. In much the same way that discussions of special relativity almost inevitably involve trains, discussions of general relativity and the equivalence principle almost always begin with elevators. An elevator provides a relatively familiar context that captures the two essential features needed to understand general relativity: an acceleration that's experienced as a change of gravity and the restriction of observations to a small local region.

That second qualification is key to using the equivalence principle in practice because it's critical to distinguish between what you see happening in your immediate area and what's happening on a global scale. As we learned from our discussion of special relativity in Chapter 12, it's impossible to talk about measuring events that occur at different places without talking about how you synchronize those measurements. General relativity complicates this further by making measurements of time and space depend not only on how you're moving but where you are relative to a massive object.

For practical discussions of the equivalence principle, it's useful to split it into two parallel statements; as with special relativity, these are generally phrased as prohibitions against distinguishing between particular types of motion:

1. Physics according to an observer who is freely falling under the influence of gravity is indistinguishable from physics according to an observer who is at rest in a world without gravity; and
2. Physics according to an observer who is at rest in a gravitational field is indistinguishable from physics according to an observer who is accelerating in a world without gravity.

In both of these cases, the observations being discussed are *local* observations: that is, there's no experiment you can do *inside an elevator* that will allow you to distinguish between gravity and acceleration.

An elevator is also a good thought experiment for exploring the equivalence principle because most people who have ridden in one have directly experienced the apparent change in gravity. When an elevator starts into motion, you briefly feel a change in weight: if it's headed up, you temporarily feel heavier, while as it starts down, there's a brief moment of less weight. The same thing happens as it comes to a stop at a floor, but in reverse: if it's heading up, there's a moment of lowered weight as it slows to a stop, while if it's moving down, there's a moment of increased weight. If for some reason you were standing on a scale in the elevator (another of those things that only people in physics problems do), you would see these changes in weight reflected in the scale readings.*

The lesson of the equivalence principle is that we can take this to the extreme case, where gravity can be entirely absent. To return to Einstein's original daydream, an elevator in free fall would produce a situation that's completely indistinguishable from gravity ceasing to work at all. A passenger inside the elevator would feel weightless, and any object, like a ball, released as the elevator began to fall would simply float in the air, holding the same position relative to the walls. The results of any experiment done inside the elevator will look for all the world as if the elevator is somehow floating in interstellar space, far from any other objects.

An observer outside the elevator, taking a more global view of the situation, would say this happens not because gravity has shut off but because gravity affects everything equally. The elevator, the passenger, and the dropped ball are not maintaining constant positions in an absolute sense. They only hold position *relative to one another* because in fact they're all falling together at exactly the same rate. This is exactly the same mechanism that produces the microgravity environment experienced by astronauts on the International Space Station: the earth's gravitational pull on the station is only about 10 percent less than it would be at ground level, but all the objects on the station are accelerating

---

* The exact change depends on the elevator and the person, but for an adult with a mass of 60 to 70 kilograms, the scale reading would likely change by three or four kilograms (some six to eight pounds) during acceleration.

A passenger playing with a ball inside an elevator experiences a world without gravity:
the ball just hangs in midair. An outside observer, however, sees this as a result of
gravity, as the ball, the passenger, and the entire elevator are freely falling. (Cartoon
people in the figures in this chapter are from drawings by Claire Orzel.)

toward the earth at the exact same rate, and thus their relative positions don't change.*

The reverse situation can also be made to work: an elevator that *really is* out in space far from anything else can be made indistinguishable from an elevator at rest near the surface of the earth. A passenger inside either elevator can feel the normal sensation of weight; an object dropped will fall to the floor with the usual 9.8 m/s^2 acceleration, and so on.

How can this be achieved? By accelerating the elevator in the "upward" direction at 9.8 m/s^2. In this case, an observer outside the elevator would say that the passenger's sensation of weight comes not from gravity pulling down, but from the floor pushing up on their feet to move them with the elevator.

---

* They don't *hit* the earth because they're also moving sideways at 4.7 miles per second. The acceleration caused by the earth's gravity only changes the direction by enough to bend the orbit into a (nearly circular) elliptical path around the earth.

A passenger inside an elevator experiences normal gravity, with a dropped ball accelerating toward the floor. An outside observer sees this instead as a result of acceleration of the elevator: the ball is holding still, while the floor accelerates up toward it.

The dropped ball is not truly falling: it's remaining fixed in place relative to the outside observer, but the floor is rushing up to meet it.*

The core of the equivalence principle is this inability to distinguish between gravity and acceleration. Whatever the passenger feels inside the elevator—whether gravity is present or entirely absent—can be explained by an observer in the opposite situation by the proper choice of acceleration. And just as with the original principle of relativity—The Laws of Physics Do Not Depend on How You're Moving—this has to extend to the behavior of light.

---

* A similar effect can be provided by rotation, for example, by swinging the elevator in a large circle by the cable attached to the top of the car. In this case the "gravity" felt inside the elevator is seen by the outside observer as the result of the centripetal acceleration of the floor toward the center of rotation. Simulating gravity by rapidly spinning a space station is a staple of both science fiction stories and plans for long-term human space exploration.

BENDING THE SHORTEST PATH

The most spectacular change to the behavior of light predicted by general relativity is the result that got the *New York Times* headline writer so worked up in 1919: Light bends under the influence of gravity. While this is something that can only directly be seen in astronomy, we can understand the idea by again thinking about elevators.

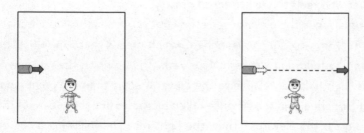

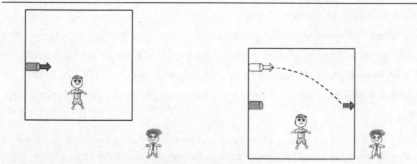

A weightless passenger inside an elevator sees a laser beam travel straight across the car, while an outside observer sees the light follow a curved path as the elevator falls.

In this case, we should imagine a passenger who is weightless in an elevator shining a laser horizontally across the elevator car. That passenger will see the light follow a perfectly straight path across the car, hitting the far wall at precisely the same distance above the floor as its starting height. There's nothing else it *can* do, as light follows the shortest path between two points, which in this case is a straight line, more or less by definition.*

---

* Going beyond the subject of this book, nearly all of classical optics—mirrors, prisms, lenses, etc.—can be derived from this principle, that the path followed by a ray of light is the one that takes the least time. This principle was first stated by Pierre de Fermat in 1662 when

The equivalence principle, though, tells us that the weightlessness experienced inside the elevator must also be explainable as the consequence of acceleration. Specifically, the passenger inside the elevator will think that gravity is nonexistent if the elevator is in free fall, dropping straight down at the full acceleration of gravity. This poses a problem for the behavior of light, because while the time required for light to cross the elevator is very short, it's *not* zero, and during that time, the elevator will have fallen some distance as seen by the outside observer who feels the full effect of gravity.

The passenger *inside* the elevator sees the light hit the far wall at precisely the same height from which it was launched, which means that to the outside observer, it can't be following a straight-line path. The point on the wall where the light hits is slightly lower relative to the elevator *shaft* than the point from which the light launched, and the distance fallen increases like the square of the time it took to cross the elevator.* Thus, the light must be following a curved path, tracing out a parabolic arc according to the outside observer in order for the inside observer to see it as straight.

The effect of gravity bending light is tiny for even a mass as large as the earth, but given a big enough object, it should be detectable, and that's what the 1919 Eddington expedition set out to do.† They photographed stars near the rim of the sun during a total eclipse, and then they compared the positions of those

---

discussing the bending of light by glass as a consequence of the change in speed. The same basic idea, that the trajectory followed by an object is one that minimizes some quantity associated with the motion, can be applied to other domains of physics—in classical physics this is the "Lagrangian formulation" and in quantum physics it's the "Feynman path integral formulation." While this may seem like a lot of additional work, for certain kinds of problems, the resulting equations are much easier to solve, making this an incredibly useful technique.

* That is, if we instead measured the position of the beam halfway across the elevator, we would see that it had dropped by one-fourth the distance relative to the shaft.

† Eddington's team was not the first to attempt this. Einstein had predicted the bending of light as early as 1911, and several expeditions had been launched to try to observe this in the years between that first prediction and the 1919 success. All of them met with misfortunes: bad weather, malfunctioning equipment, and in one case the outbreak of World War I, which landed a team of German astronomers in a Russian POW camp for some time. The various delays were for the best, as it turned out, because the 1911 prediction was based on an incomplete version of the theory and predicted a value of the deflection that was half as big as it should be. The 1919 images showed a value that matched well to the correct prediction from the completed theory of general relativity, which secured Einstein's fame.

stars when their light had to pass close to the sun to the positions of the same stars at another time, when the sun was not there. They showed that stars near the rim of the sun appeared to be pushed slightly outward, exactly as predicted by Einstein's theory.

The bending of light by massive objects has been reconfirmed numerous times since the Eddington result of 1919, during other eclipses and using other massive objects. The most spectacular example is the 2019 Event Horizon Telescope image of a black hole with a mass of four million times the mass of the sun located at the center of the M87 galaxy. The black hole looks like a "shadow" in the image, because its gravitational pull is so intense that light that should have come to us from that region is either trapped inside, unable to escape, or bent out of our line of sight by gravity. The size and shape of that shadow can be matched to theoretical models using general relativity, which astronomers can then use to determine the properties of the black hole itself.

The bending of light points to the reason why general relativity is usually described in terms of "warping space and time." Following the principle that light always takes the shortest path between two points, this suggests that the shortest path is no longer a straight line but a curve, at least as seen by the outside observer. This seems like a strange idea, but in fact it has a familiar analogue: if you look at the route followed by transcontinental flights on a flat map, you'll see that they follow curved paths—a flight from San Francisco to Paris loops up over Canada and Greenland. This happens not because airlines are run by idiots, but because the map is a flat projection of a spherical Earth: the "curved" route *really is* the shortest possible path, once you take into account the true shape of the earth.*

Light bends because geometry in the presence of massive objects is not the flat geometry we're familiar with from drawing small shapes on pieces of paper, but something more akin to the geometry of extremely large paths on the curved surface of the earth. What matters for the path of light is also not the geometry of space alone, but *space-time*: It's not just a path through space from one point to another but a trip through *time* from the past into the future. As Hermann

---

* The shortest route between two points on the surface of a sphere is one that's a piece of a "great circle," a route that passes through both points as it makes a loop around the full circumference of the sphere back to the start.

Minkowski pointed out in 1908 (quoted at the end of Chapter 12), space and time are inseparable and mixed together in different ways for different observers.

What we see as the effect of gravity is reflecting a change in the nature of space-time itself, changing the shape of the "shortest path" between two points. To a distant observer, this looks like a change in the precise mix of what is space and what is time at different locations, with "move forward into the future" taking on a little "move closer to the massive object" character. The American physicist John Archibald Wheeler famously and pithily summarized the theory as, "Space-time tells matter how to move; matter tells space-time how to curve."

This geometric interpretation also explains the 10-year gap between special relativity in 1905 and general relativity in 1915. While Einstein was not immediately fond of Minkowski's description of his theory in geometric terms, as he began working through the consequences of the equivalence between gravity and acceleration, he realized that the geometric description was essential. The proper expression of his ideas is in terms of a curved four-dimensional space-time, but that was not a branch of mathematics commonly taught to physicists in Einstein's day. As a result, he spent years laboriously learning (with the help of his longtime friend Marcel Grossmann, among others) the mathematics of four-dimensional geometry in curved spaces.* This was not a smooth process, and when Einstein finally completed the theory in late 1915 and calculated one of the results that would become crucial supporting evidence for the theory,† he became so excited he had heart palpitations and wound up spending several days in bed.

---

* Much of the mathematical apparatus he needed had in fact been invented decades earlier (the general topic is often called "Riemannian geometry" after the seminal work of Bernhard Riemann in the 1850s). This nearly led to Einstein being scooped on his own theory by the mathematician David Hilbert of Göttingen University, who knew the necessary techniques and quickly put together his own version of the theory. Hilbert famously quipped that "Every boy in the streets of Göttingen understands more about four-dimensional geometry than Einstein," but that the theory would not have been possible without Einstein's initial insight about the physics involved.

† The perihelion shift of Mercury, which like the moon is in an elliptical orbit that changes orientation over time. The rate at which the orbit changes is faster than can be explained using Newtonian gravity, which led lots of people to look for another planet even closer to the sun that might explain the perturbation. Einstein's completed theory gets this shift almost exactly right and was one of the best pieces of evidence supporting it before the Eddington eclipse results.

GRAVITATIONAL REDSHIFT

General relativity is one of the great intellectual triumphs in the history of physics, but to this point it might not seem like it's particularly relevant to timekeeping. In fact, though, there's a very deep connection between the two, which we can understand by once again thinking about experiments in elevators. In this case, we think about a passenger playing with a laser and ask what effect gravity has on the frequency of the light.

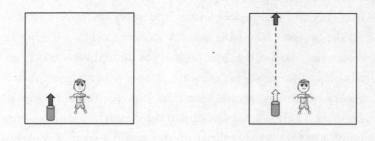

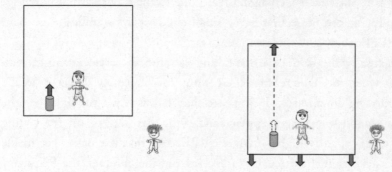

A weightless passenger inside an elevator sees a laser sent from the floor to the ceiling arrive with the same frequency as it left, but an outside observer would say they *should* see the light Doppler shifted by the increasing speed of the elevator. Thus, the light must be redshifted by gravity to compensate for the Doppler shift.

This time imagine our weightless passenger in an elevator with a laser pointer mounted on the floor shooting at the ceiling. We envision them not only measuring the perfectly vertical straight-line path followed by the light but also the color of the light as it hits the ceiling. In the absence of gravity, nothing should change this, so the frequency of the light reaching the ceiling should exactly match the frequency of the light that left the laser.

Seen by the outside observer, though, that result is mildly surprising, because again, the elevator has fallen some distance. Thanks to the acceleration of gravity, the elevator is moving slightly faster when the light reaches the ceiling than it was when the light left the floor. In which case, the frequency seen at the top of the elevator ought to appear slightly higher than the frequency of the light that left the laser because of the Doppler effect.

The Doppler effect is a change in the frequency of waves when the observer measuring the frequency and the source emitting the waves are moving relative to one another. If the observer is moving toward the source, they see a higher frequency: they're moving toward the next peak at the same time it is moving toward them, so the time between peaks decreases. If they're moving away, they see a lower frequency because it takes a little extra time for the next peak of the wave to reach them.* This is most familiar in the context of sound waves: the engine sound of a passing car is shifted up in frequency as it approaches and down as it moves away, changing from one to the other very rapidly as it passes, producing the noise that every small child knows signifies a race car: EEEEEEEEEEooooowwwwwwww . . .

In the case of the laser in a free-falling elevator, the accelerating motion means that when the light reaches the ceiling, the frequency detector on the ceiling is moving downward a little faster than it was when the light left the floor. That should lead to a Doppler shift, with the detector on the ceiling recording a very slightly higher frequency. The fact that the passenger inside *doesn't* see such a shift must mean that, to the outside observer, the frequency of the light has shifted downward to compensate. For visible light, this means a shift toward the red end of the spectrum, so this is commonly called a "gravitational redshift."

If we reverse the direction of the light, a similar argument holds. The floor is moving away from the light slightly faster when the light arrives than when it

---

* The moving observer would also measure a correspondingly shorter wavelength, with the waves appearing to squish down or stretch up as they catch up to or run away from the next peak.

left the ceiling, so the downward-moving light must be shifted *up* in frequency to compensate: a "gravitational blueshift."

The shift involved is tiny but has been confirmed by experimental measurements. The first and most famous test was done by Robert Pound and Glen A. Rebka Jr. in 1959, using gamma rays sent up and down a "tower" in the physics building at Harvard.* They placed a gamma-ray source at the top of the tower, then put a thin film of material in front of their detector at the bottom that should absorb gamma rays of that frequency. They mounted the source on the paper cone of a loudspeaker so they could move it back and forth at different speeds, then determined the frequency shift by measuring the velocity needed for the absorber to see the gamma rays from the source Doppler shifted by enough to maximize the absorption. They found that gamma rays sent up the tower shifted down in frequency (and those sent down the tower shifted up, when the source and detector switched places) by exactly the amount predicted.

The Pound-Rebka experiment raises an interesting question, because the two scientists were not in the exact situation of the two observers in our elevator thought experiment. The source and detector at the top and bottom of the tower were fixed in place, neither moving with respect to the other (other than the tiny vibration of the speaker). If they brought their sources and detectors together on some middle floor, they would be absolutely identical, and neither experimenter would observe any change in their apparatus as it changed elevation. So how did they explain the difference in frequency measured by their partner at the other end of the tower?

To put it in concrete terms, imagine Pound on the top floor of a tall tower with a source of light that oscillates 1,000,000,000,000,000 times per second; when the light arrives at the bottom, Rebka records the frequency as 1,000,000,000,000,010 oscillations per second.† From Pound's perspective, though, nothing changed with his source or the light it sends out, so why should Rebka see a difference?

The answer to the dilemma is that measuring a frequency implicitly requires a clock: to measure a number of oscillations per second, you need a way to

---

* Essentially, just a stairwell stretching from the basement to the attic.
† This is a lower frequency and a larger fractional shift than what they actually measured, but it makes for nice round numbers.

count one second. And the same mixing of space and time that leads to the bending of light makes clocks at different altitudes tick at different rates. What looks like purely a step one second into the future (with no motion in space) for one observer will seem to the other observer at a different altitude to be a slightly different step in time (plus a little motion in space).

Thanks to this shifting mix of space and time, each observer will see the other's clock moving through time at a slightly different rate, similar to what we saw in Chapter 12 where each train sees the other's clocks running slow. In general relativity, though, the change is not symmetric. The gravitational shift depends not on their relative motion but their position relative to the earth. The end result from a global perspective is that the clock at ground level is ticking more slowly.

From Pound's vantage point atop the tower, Rebka's clock is ticking too slowly: 1 second on Rebka's clock is 1.00000000000001 seconds on Pound's, which is why an extra 10 oscillations fit into each "second" at ground level. This also explains why the frequency of the light sent up the tower is too low—if the number of oscillations in a too-long "second" at ground level is the same 1,000,000,000,000,000 that Pound sees from his source, that means Rebka's source must be putting out light at 999,999,999,999,990 oscillations per second.

The same argument applies from the ground-level perspective, but in reverse: from Rebka's perspective, his clock is correct and his light has the proper frequency, while Pound's clock is ticking a bit too fast. The too-short "second" at high altitude accounts for Pound measuring Rebka's light as too low in frequency, and it explains why Pound is sending down light of too high a frequency.

This "gravitational time dilation" is the final prediction we can make from the equivalence principle, and the one most directly relevant to timekeeping. Clocks at different altitudes—that is, different distances from an object massive enough to produce a significant gravitational pull—will necessarily tick at different rates. This idea was used for dramatic effect in Christopher Nolan's 2014 science fiction epic *Interstellar*, when decades pass on Earth as the protagonists make a brief visit to a planet orbiting close to a supermassive black hole. In the real world, the shift is nowhere near so dramatic, but it exists and creates both problems and opportunities for scientists operating ultra-high-precision clocks.

TIMEKEEPING AND GRAVITY

The effect of Earth's gravity on the ticking of a clock is extremely small, amounting to a bit less than 10 nanoseconds per day per kilometer above sea level. This is, however, well within the range of sensitivity of a cesium atomic clock. As noted in the previous chapter, the clocks in the GPS constellation are adjusted to correct for the gravitational time dilation at the 20,000 kilometer altitude of the satellites.

The shift for Earth-based laboratories is smaller, but it still must be taken into account by standards laboratories in putting together the time for UTC. For most labs, this is a small correction, but for the US National Institute of Standards and Technology in Boulder, Colorado (at an elevation of 1.62 kilometers above sea level), the shift is fairly substantial. In 2016, a team of undergraduate students compared a commercial atomic clock to the corrected-to-sea-level time broadcast by GPS at a number of different locations in Colorado at elevations from 1,300 to 4,300 meters, and were able to clearly detect the gravitational shift.*

While the gravitational time dilation poses a bit of a problem for metrologists constructing the global timescale, this can be turned around and made a tool for other sciences. In much the same way that Jean Richer's seconds pendulum helped reveal the true shape of the earth back in the 1670s, comparing clocks at different locations can be a tool for geodesy, helping to reveal differences in gravity on a much smaller scale. Networks of clocks on the earth and in space can also be used to search for exotic physics, both in the form of possible deviations from the predictions of general relativity and also in more transient effects caused by the passage of gravitational waves.

All of these applications, though, require greater timing precision than is possible even with state-of-the-art cesium clocks. The best cesium clocks are good to a few parts in 10^{16}, which corresponds to an elevation change of around a meter. Applications in geodesy and fundamental physics would need clocks that are better by at least a factor of 100, which is not currently feasible with cesium. There are, however, experimental clocks under development that reach

---

* M. Shane Burns et al.,"Measurement of gravitational time dilation: An undergraduate research project," *American Journal of Physics* 85, 757 (2017), https://doi.org/10.1119/1.5000802.

this level of precision using other chemical elements and new measurement approaches. In Chapter 16, we'll discuss how these new tools ensure a bright future for timekeeping.

First, though, let's examine a class of technologies that developed in parallel with quantum mechanics and relativity. While they never expanded the frontiers of physics, the industrial manufacture of mechanical watches and the rise of quartz timekeeping did something that was transformative in its own way: they brought accurate timekeeping to the masses.

Chapter 15

TIME ENOUGH FOR EVERYONE

I have two wristwatches that I keep for different occasions. Both of them look very similar—analog dial, brown leather band—but their mechanisms are very different. One is a fancy Omega self-winding mechanical watch from the 1960s, passed on from a dear friend of the family, that I like to wear on special occasions. The other is a modern Seiko battery-powered watch for everyday use,* and I've probably spent more on replacement bands and batteries in the 10 years I've owned it than it originally cost.

For the purposes of this book, the most interesting thing about my two watches is that the cheap one keeps better time. This is not because of any particular defect in the mechanical watch—on the contrary, it's a high-quality piece of precision engineering, which is why it still runs more than 50 years after it was made. The issue is technological in origin: the mechanical watch uses elements whose general form would've been familiar to John Harrison back in the 1700s—balance springs, mechanical escapements, etc.—though many of those elements have been improved since his day. The battery-powered one, on the other hand, keeps time based on an entirely different kind of motion: tiny vibrations of a miniature "tuning fork" made of quartz.

---

* Well, every day when there isn't a pandemic that keeps me from leaving the house, as there was for several months as I wrote this in 2020.

A quartz crystal may sound like an exotic element to include in a watch, a bit like Harrison's use of diamond pallets to reduce friction. Unlike those gems, though, quartz is an extremely common mineral and has become ubiquitous in timekeeping—unless you specifically seek out some other sort of mechanism, almost any electronic clock you buy these days will contain a quartz crystal as a regulator. This, in turn, has enabled an incredible democratization of timekeeping: where in the past, a precision clock or watch was costly and rare, an heirloom item, in the twenty-first century the relationship between precision and cost has reversed. These days, high-precision quartz timekeepers are cheap and common, practically disposable, while purely mechanical clocks and watches with lower intrinsic accuracy are luxury items.

MAKING THE DOLLAR FAMOUS

The first mechanical watches, as we saw in Chapter 10, were hugely expensive, particularly those intended for use at sea. As the demand for them increased, pressure to produce ever-cheaper versions also increased, but watchmaking remained a very fiddly business through much of the nineteenth century.

Mass production entered the field of watchmaking very slowly, enabled by advances in materials and manufacturing techniques. Improvements in metallurgy played a key role, allowing watches to be made from more durable steel rather than brass. Steel parts could be produced more cheaply and reliably than brass ones and would stand up to greater abuse. True precision devices, though, continued to use the techniques pioneered by Harrison, including low-friction jewels at key points in the works.

The idea of standardized mass manufacture was slower to take hold, even as the Industrial Revolution began in earnest, because the degree of precision required for reliable timekeeping is so extreme. A watch needs to tick 86,400 times a day, every day, to keep accurate time, so it doesn't take long for small imperfections in manufacture to add up. An error of one part in 10,000—one one-hundredth of a percent—in the period of the balance spring throws a watch off by about a minute per week, a level that begins to be annoyingly obvious. The traditional means of dealing with this was to manufacture the balance wheel and spring as a pair, tailoring the strength of the spring to match the mass and size of the wheel. These were individually adjusted to achieve the desired accuracy, a process that was as much art as science, and very time-consuming. Jewels to reduce friction further complicated matters, requiring each piece to be individually cut and fitted securely in the appropriate position.

Large-scale manufacturing of watches for mass consumption using recognizably modern techniques began in France in the late 1700s but soon shifted across the border to Switzerland, particularly in the villages of the Jura region. The general shape and layouts of watch components were standardized, and rough pieces were turned out en masse in factories. These pieces were then distributed to numerous cottage shops for fine finishing and the process of assembly, testing, and adjusting. This distributed manufacturing involved whole families and multiple shops, with some estimates suggesting as many as 150 people involved in making and assembling the parts for a single watch.* While the introduction of machine tools pushed the industry toward standardized and interchangeable parts, the manufacturing tolerances possible with the technology of the day did not allow true interchangeability.

The crucial innovation that enabled watchmaking on an industrial scale came from the Waltham Watch Company in the United States, who realized in the 1850s that rather than individually crafting watch parts to have the perfect match of spring and wheel from the start, they could simply make parts in bulk and match them afterward. They weren't able to achieve perfect standardization with the available manufacturing techniques, but once the parts were made, it was relatively simple to measure their properties to high precision and sort them into different groups. This sorting dramatically simplified the process of matching spring to wheel: when assembling a watch, one could simply pair a wheel that had been measured to be a little heavier than average with a spring that had been measured to be a little stronger than average, or a lighter wheel with a weaker spring. The raw parts output by the machines weren't truly interchangeable in the sense of being able to pair *any* wheel that came off the line with *any* spring, but quality control measurements were good enough to ensure

---

* This figure, like most of the numbers cited in this chapter, comes from David Landes's magisterial history *Revolution in Time* (first published by Belknap Press in 1983).

that a wheel and a spring chosen from matching bins would keep accurate time with far less need for individual adjustment than in a traditional watch.*

The quality of these watches caught the Swiss by surprise—David Landes describes Edouard Favre-Perret, a Swiss industry representative at the centennial exhibition in Philadelphia in 1876 marveling at the quality of the American product. A randomly selected Waltham watch from their fifth quality rating, with no individual adjustment, kept time to within 30 seconds over six days. Favre-Perret quoted a Swiss watch adjuster declaring that "one would not find one such watch in fifty thousand of our manufacture."

With a bit of rapid retooling, and thanks in part to their large existing corps of skilled artisans, the Swiss were able to adopt and adapt the American innovations, retaining their position as the preeminent watchmakers in the world. They could turn out watches that were competitive in price while also being thinner, lighter, and more accurate than the competition. By the end of the century, quality mechanical watches were being made in huge numbers on both sides of the Atlantic. In the middle 1800s, relatively large watch manufacturers were making tens of thousands of watches a year, and in 1870 the total output of the Jura region was around a million watches. In 1896, the Ingersoll brand introduced their Yankee model, made by the Waterbury Watch Company, which was the first true "dollar watch" (its one-dollar price was about a day's wages for an ordinary laborer). In 1898, Ingersoll alone sold a million of these; by the end of World War I, they had sold tens of millions in the United States and Britain.

The "watch that made the dollar famous" (a later ad slogan from Ingersoll) was a stripped-down product compared to a true precision timekeeper—it used a simplified escapement and no jewels on the contact surfaces—but it was good enough. These watches might lose tens of seconds over a few weeks of operation, which would be disastrous for a ship at sea but posed no great problem for a worker who simply needed to get to a meeting on time. Reasonably accurate personal time was within easy reach of essentially anyone who cared to have a watch.

---

* It is important to note that the final assembly and testing was still done by skilled watchmakers, with some individual tuning to ensure that the watch kept good time. The process was extremely streamlined, though, and even without the individualized adjustment, watches worked well enough for most people.

CHRONOMETERS AND COMPETITIONS

In parallel with the rise of industrial-scale manufacture of mechanical watches, there was a second, more exclusive branch of the industry, focused on precision and reliability. The Omega watch I mentioned at the start of this chapter comes out of this tradition. This strand of the story is perhaps best illustrated by the rise and fall of international chronometry competitions.

These competitions arose out of the need to certify the performance of individual watches as sufficiently accurate for use at sea. The first chronometer evaluations were conducted simply by depositing a watch with an astronomi-cal observatory for some period during which the staff would carry the watch around and check it against solar noon. Watches that performed to the stan-dards needed for navigational use would be certified as having met the stan-dard and could then command a higher price from naval or merchant sailors. These trials became both more public and more complicated over the years, and by the 1870s there were regular international competitions sponsored by observatories at Geneva and Neuchâtel in Switzerland and Kew in the United Kingdom.

Timepieces submitted to the competition would undergo rigorous evalu-ation for an extended period under precisely defined terms—in its final years, the Neuchâtel competition lasted 45 days, with the watches mounted in five different orientations and subjected to two different controlled temperatures. Each watch entered would receive a score based on its performance, with the rankings published and heavily used in advertising by the winners.

Performance in these competitions could make and break corporate rep-utations, and they also provide us with a good way to chart the rise and fall of national-level industries. Many of the key innovations in the watchmaking industry were first developed in Britain by Harrison, Arnold, Earnshaw, and their heirs, and the British held a commanding lead in chronometry compe-titions for most of the 1800s. By the turn of the century, though, the Swiss watchmaking industry had risen to preeminence, winning first prize in the competition at Kew for the first time in 1903, with a then record score of 94.9 out of a possible 100. Several decades later, the Neuchâtel competition would help announce a new player in the industry, with four watches by Seiko/Daini placing in the top 10 in 1967.

The Swiss thoroughly dominated these competitions through the twentieth century, with a Swiss entry winning every year at Kew starting in 1907, and sweeping the top spots in their own competitions most years. These watches are mechanical marvels, with jeweled works, intricate escapements, and ingenious engineering to protect against mechanical shocks and variations in temperature. The annual competitions were a spur to innovation to the very end. The Neuchâtel competitions issued scores based on error, with zero indicating perfection: In 1955, the winning watch rated a 4.4. By 1965, this was cut nearly in half, to 2.33, and in 1967, the winner had an impressive 1.73.*

The 1967 Neuchâtel competition was the final run of this storied institution, though, because the world of precision chronometry was upended by a new technological innovation. The Omega with the top score for a mechanical watch in that year was an absolute masterpiece of engineering, but it was blown away by another entrant, which scored an incredible 0.152. Quartz watches had arrived, and the industry would never be the same.

THE QUARTZ REVOLUTION

The first laboratory clock based on a quartz crystal oscillator was built at Bell Labs in 1927 by Warren A. Marrison and Joseph W. Horton, and by the late 1930s, quartz clocks were becoming essential equipment at standards laboratories and observatories around the world. In 1939, AT&T unveiled a quartz-based display clock at their headquarters in Manhattan that was hailed as "the most accurate public clock in the world."† While the new technology caught on quickly with the scientific community, it was slower to come into

---

* The internet being the weird and wonderful place it is, there is an Observatory Chronometer Database (http://www.observatory.watch/) where you can look up the Neuchâtel competition scores for all of the mechanical watches entered between 1945 and 1967.

† A 1939 story in the *New York Times* about the dedication included an acknowledgment that the official US time is determined by the US Naval Observatory, and lots of public clocks are synchronized daily with this. The president of AT&T bragged, though, that "in the intervals between these checks with their observations, the clock in our window varies less from official time than any public clock anywhere." The *New York Times* also drily noted that while the dedication was due to start at noon, the actual start was delayed 17 seconds "owing to a mishap during the unveiling."

CRYSTAL TIME

The basic principle of a quartz timekeeper is similar in concept to the pendulum and the balance spring, in that its "tick" comes from the regular oscillations of a manufactured physical object. The central element is a small crystal of quartz cut in the shape of a tuning fork, which vibrates at a characteristic frequency determined by its shape and material properties. Unlike a mechanical watch, though, these oscillations are both recorded and initiated by electrical signals.

This grows out of a technique developed in the 1860s, which used the audible-frequency vibrations of a metal tuning fork as the basis for timing relatively fast motions. These were distinctive in that they relied on the new technology of electricity: the vibration of the fork was sustained by small "kicks" delivered by an electromagnet near the end of one of the prongs, and the frequency was determined by using a different magnet to produce an electrical signal that could drive a motor.

With the development of vacuum-tube amplifiers in the early 1900s, these components could be combined to produce extremely stable oscillations: the signal picked up from the vibrating fork is in turn used to generate the series of kicks that sustain the vibration. This feedback system will pick out and amplify a particular frequency of oscillation, in an actually useful version of the process that leads to the characteristic screeching noise when a microphone in a PA system is pointed toward the speaker.

With properly designed circuits, such a system produces an extremely stable oscillation at one very particular frequency characteristic of the fork itself. Vacuum-tube-driven oscillating forks in the 1920s and '30s performed at a level comparable with Shortt pendulum clocks, but with a much higher frequency of vibration, and their electronic signals were used to time some astronomical events to very high precision. A 100-oscillation-per-second fork at Bell Labs was used to time observations of a 1925 solar eclipse at a number of cities in New York, coordinating their measurements via electrical signals on telegraph lines. In 1960, a compact version of the tuning-fork system was even built into a high-end watch, the Accutron by Bulova, which took its timing signal from a tuning fork vibrating 360 times a second.*

---

* This is an audible frequency, which led to one of the novel features of these watches:

The quartz oscillators that are so ubiquitous today use the same basic principles of physical oscillations sustained by feedback circuits but with a few significant advantages over clocks based on metal forks. These have to do with the particular structure of quartz, which is a crystalline mineral consisting of silicon and oxygen atoms (chemically, it's silicon dioxide, SiO_2) arranged in a regular repeating pattern. One of the key features is that this crystal structure makes quartz an exceptionally rigid material, which gives a tuning fork made of quartz a characteristic frequency that is both very high compared to other materials and also remarkably insensitive to perturbations.[*] Oscillators made from quartz crystals cut with the proper orientation to the crystal have very little sensitivity to changes in temperature, making them ideal as a basis for timekeeping.

More importantly, the crystal structure of quartz gives it an interesting connection between its physical and electrical properties. This effect was first discovered around 1880 by a pair of French brothers, Jacques and Pierre Curie,[†] who noticed that putting physical stress on quartz crystals caused them to develop electrical charges on the surface. "Squeezing" the crystal along one particular direction would create a voltage difference across the crystal where the top is more negative than the bottom, while stretching it would reverse the voltage. This effect can also be reversed: applying a voltage across the crystal causes it to expand or contract by the same small amount.

---

rather than the traditional ticking sound, they made a faint humming. The Accutron and its signature hum feature in an ad pitch that opened the final season of the TV show *Mad Men*.

[*] This also posed a bit of a challenge for the pioneers of quartz clocks, who had to develop electrical systems to reduce the high frequency of the quartz oscillator down to something that could drive a motor moving the hands of a clock or update the numbers on a digital display. In the 1920s, engineers devised a simple vacuum-tube circuit to cut the frequency of an oscillating voltage in half, and modern clocks use a similar circuit built into silicon chips.

[†] Pierre would later abandon his work on materials to join his wife, Marie Skłodowska Curie, on her project of studying radioactive elements. Pierre and Marie shared the 1903 Nobel Prize in Physics for this work, and Marie was awarded the 1911 Nobel Prize in Chemistry for related work; Pierre likely would have shared that as well, but he was killed in a traffic accident in 1906, and Nobel prizes are not awarded posthumously.

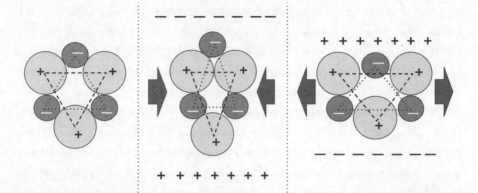

The microscopic origin of piezoelectricity. In normal quartz, the positive silicon ions and negative oxygen ions are arranged in a hexagonal lattice so that their charges balance out and the material is neutral. When squeezed from the sides, the oxygen at the top and silicon at the bottom are pushed outward, causing the top of the crystal to become negative and the bottom positive. When stretched, the charges reverse.

The Curies dubbed this phenomenon "piezoelectricity," from the Greek verb *piezein*, "to squeeze," and determined that it is associated with the crystal structure of quartz (and a whole family of other materials). We can understand this by looking at the microscopic structure of quartz crystals, which consists of negatively charged oxygen atoms and positively charged silicon atoms arranged at the corners of a hexagon (when viewed from a particular direction), as seen in the figure. In the normal state, when no stress is applied, this hexagonal crystal element can be broken up into two equilateral triangles, one with two positively charged silicon atoms at the top and a third at the bottom, the other with two negatively charged oxygen atoms at the bottom and the third at the top. This arrangement is nicely symmetric, so the effects of the charges balance each other out: an observer above the crystal looking at the top surface sees two positive charges and one negative, but the positive charges are slightly farther away, so their effect just exactly cancels the single closer negative charge.

When the crystal as a whole is placed under stress, squeezing it from the left and right sides, the atoms shift position relative to one another: the two silicon atoms at the top squeeze together and push the oxygen between them up a bit, while the two oxygens at the bottom get closer and force the silicon down slightly. The triangles distort under the strain, pushing the single charge at either end farther out, and the balance is lost: the top side of the crystal behaves like it has a negative charge, while the bottom side behaves like it has a positive charge. This is detectable as a voltage difference between the top and bottom, with the amount of voltage increasing as the squeezing force (and thus the distortion of the crystal structure) increases.

If instead of squeezing the crystal, you apply a voltage across it with the top positive and the bottom negative, this will induce a physical change in the crystal. The negative oxygen atom at the top is pulled upward, while the positive silicon atoms are pushed down, and vice versa on the bottom. As the triangles stretch in the vertical direction, the bonds between atoms pull them in to the left and right, and the crystal shrinks. If you reverse the voltage (top negative, bottom positive), you reverse the effect and the crystal expands.

Nowadays, piezoelectric materials have a wide range of technological applications, as pressure sensors (measuring the voltage generated as a force is applied) and as actuators for moving objects small distances. The audio speakers found in cell phones and earbuds make noise by driving vibrations in small piezoelectric crystals to create sound waves. They are even found in the triggers of grill lighters, which work by squeezing a chunk of piezoelectric material so hard that the high voltage induced makes a spark in the air gap between two wires connected to the different faces of the crystal.

In timekeeping, the piezoelectric properties of quartz are used for both driving and detecting the oscillations. Electrical impulses are used to stretch and compress the quartz fork, which starts it oscillating; that oscillation in turn creates piezoelectric voltages that are picked up, amplified, and used to keep time. A feedback arrangement similar to the one described above for keeping a tuning fork oscillating keeps the quartz oscillator in motion, providing an exceptionally stable time signal. The standard quartz oscillators used in clocks and watches are cut to oscillate 32,768 times per second. This particular value is chosen because it's 2 to the 15th power, so a simple chain of (relatively easy to make) electronic divide-by-two elements can bring the frequency down to the one-per-second rate needed to drive a clock display.

widespread consumer use. Early quartz clocks required vacuum-tube amplifiers to provide the necessary feedback, making them bulky and inconvenient until the development of transistors and integrated circuits in the 1960s.

As portable quartz clocks began to be practical, they were submitted to the official chronometry competitions held in Switzerland and far outdid the mechanical timekeepers. The first entries in 1961 were marine chronometers that were just barely practical, with batteries that needed changing weekly. One of them scored a 1.2 at Neuchâtel, though, handily beating the 2.8 of the best mechanical chronometer. By 1967, the top quartz entrant in the chronometer category was considerably smaller and less power hungry, and also a hundred times better, with a score of 0.0099. And, as noted above, quartz timekeepers were beginning to be made in watch sizes (albeit not yet for mass production).

On Christmas Day in 1969, Seiko unveiled the world's first quartz wrist-watch for the consumer market, the Astron. While this was a significant technological innovation, offering timekeeping accurate to within a few seconds per month (comparable to the very best mechanical watches), it came with a 450,000 yen price tag,* so this was not exactly a mass consumer item. It kicked off a decade-long process of innovation, though, that completely transformed the consumer market for timekeeping. The fully electrical nature of quartz clocks led to the development of digital displays (the first, in 1972, used LEDs that required the wearer to push a button to display the time; by the end of the 1970s, these had been almost completely replaced with low-power LCD displays). Meanwhile, the rapidly shrinking size and increasing power of digital electronics allowed a dizzying proliferation of features: multiple alarms, stopwatches and countdown timers, watches with built-in calculators. And all of these kept and even improved the high standard of performance enabled by quartz crystal oscillators.

By 1977, Seiko alone was selling tens of millions of quartz watches a year, and the global market was in the hundreds of millions. By 2015, the Japan Clock and Watch Association estimated the total global market for timepieces at 1.46 *billion* watches, with 97 percent of those based on quartz oscillators.

---

* Currency conversions at 50 years' remove are always risky, but that would be around 3.5 million yen today, which is more than $30,000: then and now, it cost about as much as a new car from Toyota.

The explosive growth of quartz watches was devastating for the precision mechanical watch industry, particularly in Switzerland, where watchmakers were slow to adapt to the new technology. What remained shifted almost entirely to serve a luxury market, selling mechanical watches less as precision timekeepers than status symbols. A marketing slogan adopted in the late 1970s captured the spirit of the shift well:

> *A Patek Philippe* [watch] *doesn't just tell you the time, it tells you something about yourself.* *

THE DEMOCRATIZATION OF TIME

The development of mass-produced quartz watches brings together the two threads of the watchmaking story into a single device. For most of the industry's history, watches could compete on either price or precision, but it was rare to find a watch that offered both highly accurate time and a low price. Low-price mechanical watches were generally just "good enough"—not to the standards of an observatory-certified chronometer, but in an age of common public clocks, it's no great sacrifice to need to reset your watch every week or two. These watches were sold on the basis of other virtues: Timex, for example, was famous for its slogan "Takes a licking and keeps on ticking." Their ads showcased the durability of their watches, demonstrating that they would continue running despite being attached to the hooves of horses, the propeller of a boat, or the bat of a baseball star.

Quartz fundamentally changed the business, making incredible precision cheap enough to be essentially disposable. Nowadays, anybody who wants one can readily obtain a quartz watch whose performance is 10 times better than the finest mechanical watches ever made.

While the "quartz crisis" stands out as a singularly devastating moment for the Swiss watch industry, in a much deeper way, it's very much of a piece with the whole long history of timekeeping that we've been tracing in this book.

---

* Quoted in Landes; images of these ads are readily available online.

Over the last 5,000 years of human civilization, the science and technology of tracking time has shifted bases repeatedly, each development improving its accuracy. And each leap forward in accuracy has been accompanied by an increase in the availability of timekeeping for ordinary people.

In the earliest societies, like those of the Neolithic people of Britain, timekeeping was a monumental but elite pursuit. The Newgrange passage tomb is a marvel of large-scale engineering that must have marshalled the labor of thousands of people, but the signature "tick" of this massive clock would only be visible to the tiny number of people who can fit in the central chamber, accessible only by invitation. Stonehenge is more open and dramatic, but again, fixed to a particular location and thus only visible to pilgrims able and willing to travel there.

More elaborate astronomical calendar systems, like those of the Egyptian and Mayan civilizations, can track the seasons over a wider geographic range, but these retained an elite, divinatory purpose. Schemes for predicting the dates of eclipses or the rising of Venus were used to bolster political and religious power. As civilization spread more widely, luni-solar calendars like the Hebrew, Julian, and eventually Gregorian calendars came into use, allowing anyone who can count days to predict the seasons and important holidays with reasonable accuracy.*

On shorter timescales, the development of water clocks like Su Song's tower clock and the mechanical cathedral clocks of early modern Europe helped make time a matter of public knowledge in urban areas. Ringing bells announce the hours to anyone within earshot, allowing a finer division of the day for a wider segment of the population. Reliable mechanical clocks and watches made navigation at sea safer and more reliable and also brought highly accurate timekeeping into a huge number of private homes. Finally, the growth of global telecommunications networks allowed people all over the world to keep their own clocks synchronized with those in major cities by telegraph and radio and eventually the internet and GPS.

In this light, the mass production and distribution of watches, first mechanical and then quartz, is just another inevitable step in a process of democratization stretching back millennia. Literally billions of people today have access to

---

* Though setting the date of Easter continued to be controlled by religious authorities.

timekeeping at a level of precision that would've seemed impossible little more than a century ago. This in turn has profoundly affected the way we live our lives, for good and ill, allowing better coordination but also tighter scheduling between people, making us all busier.

And, of course, quartz clocks and watches are themselves vulnerable to being supplanted. Whichever of my wristwatches I choose to wear on a given day, the antique mechanical one or the modern quartz one, when I want to know *exactly* what time it is, I pull out my phone. While cesium fountain clocks remain directly accessible only to a small number of research scientists, thanks to the internet and the GPS constellation of satellites, I have ready access to a close approximation of UTC almost any place I go. That level of accuracy is more than I really *need* for any of my daily activities, though it's nice to know that so long as the network is up, I have access to the correct world time. For certain applications, though, even UTC is not enough; in the next chapter, we will conclude our trip through the history of timekeeping with a look toward the future.

Chapter 16

THE FUTURE OF TIME

The GPS constellation broadcasts a time signal that is regularly synchronized with UTC(USNO), and thus can be used to set local clocks to a good approximation of atomic time. The service is also regularly used by national standards laboratories around the world to do the clock comparisons required for assembling the UTC scale. Two labs comparing their local clocks to the signal from a GPS satellite that's simultaneously visible to both of them can get the difference between their clocks to within a few nanoseconds, and this can be improved somewhat by using multiple satellites and some signal processing tricks.

In parallel with GPS, atomic time is also distributed via the internet, with multiple reference sources combined via the Network Time Protocol (NTP) system. Computers connected to the network can synchronize their local time to a realization of UTC to within a few milliseconds. In practice, home computers are often a bit worse, but the National Institute of Standards and Technology maintains a website that will tell you the difference between your computer's clock and UTC(NIST) to within a millisecond. On the rainy morning when I first typed these paragraphs, for example, my desktop computer's clock was running 0.589 seconds behind official US time.

Between NTP and GPS, people who care about accurate timing have ready access to clocks sufficient for essentially any everyday purpose, and quite a few that go beyond the mundane "What time is it?" These services provide UTC timestamps for power companies looking to keep their generators synchronized across the power grid, financial institutions tracking high-frequency trades, and even astronomers seeking to combine signals from far-flung radio telescopes to achieve higher-resolution images. The Event Horizon Telescope image of the black hole in M87 was possible in part due to time signals derived from GPS that let the researchers piece together signals from different telescopes around the world, effectively making one huge telescope. Those timestamps are traceable back to the cesium atomic clocks providing time to the GPS constellation.

Given the wide range of applications satisfied with the existing atomic time services, it may seem like the long history of progress in timekeeping, stretching back to Neolithic solstice markers, has reached its end. Cesium atomic clocks, good to within a second over millions of years, seem like the pinnacle of time-keeping technology, and all that's needed from here on out is just a program of regular maintenance.

In reality, though, cesium clocks are nowhere near the end of the time-keeping story. While the existing atomic time services are adequate for almost everything we have going on right now, physicists in standards labs around the world are hard at work on a next generation of atomic clocks. These clocks are still very much in the developmental stage, but their performance can be hundreds of times better than the best cesium clocks. It's possible that, years from now, some other atom will take cesium's place in the definition of the second, but even before that, these prototype clocks enable some astonishing measurements.

A SOLUTION FINDING PROBLEMS

The first cesium atomic clock was demonstrated in 1955, and cesium became the standard for determining the second just over a decade later, in 1967. Between those two events, though, the year 1960 saw the invention of a new technology that would open the door to atomic clocks with accuracies well beyond that of cesium: the laser.

The first working laser was built in 1960 by Theodore Maiman of Hughes laboratories, based on an idea independently developed by Charles Townes and Arthur Schawlow and a young engineer named Gordon Gould. Shortly after

WHY CESIUM?

Alongside discussing potential replacements for cesium as the basis for an atomic clock, it is worth considering why this particular element was chosen to define time. It certainly wasn't because of the appealing properties of cesium in bulk—while the metal has a pretty gold color, it has a melting point close to human body temperature (it will quite literally melt in your hand), and it explodes violently when it comes in contact with water. Even humid air can be enough to start a reaction with cesium metal, so it must be stored and handled with great care. The properties that led to cesium being chosen as the element to define time are more subtle, and partly a matter of historical accident, but they also point to issues that come into play when considering alternatives.

There are two primary features of cesium that made it attractive as an atomic time standard in the 1940–1960 time frame. First and foremost, it features a "clock transition" that has a very long lifetime. Secondly, it has a transition frequency that is relatively high (for the microwave technology available at that time). Both of these properties are desirable in any prospective replacement.

The long lifetime is essential to the process of measuring the frequency of the clock transition (described in Chapter 13), which depends on preparing the atoms in a quantum superposition of the two states and allowing it to evolve for some time. For this measurement to work properly, it is essential that this superposition is not perturbed by any external influences. This is why photos of atomic clocks always look like shiny cylinders: they are enclosed in multiple layers of metal shielding to prevent external magnetic fields from reaching the atoms. It is also important that the superposition is not perturbed by *internal* influences, particularly the tendency of atoms in high-energy states to spontaneously decay to lower energy states, emitting a photon in the process. Ideally, this spontaneous emission rate should be low enough that the lifetime of the high-energy state in the clock transition is considerably longer than the time between the interactions involved in the Ramsey interferometry measurement.

The rate of spontaneous emission for a particular atomic transition depends on a number of properties of the atom in question, the most important being the mechanism of the transition and the frequency of the light. The rate increases as the cube of the frequency, so all else being equal, doubling the frequency would increase the rate to eight times its original value (and decrease the lifetime to one-eighth of its original value). This mostly explains why transitions with frequencies that correspond to visible light tend to have lifetimes of tens of nanoseconds, while microwave-frequency transitions like the one used in cesium clocks (with a frequency something like 60,000 times lower) have lifetimes measured in years.

In addition to this frequency dependence, though, there is a factor that depends on the details of the atomic states involved. The nature of the interaction between atoms and light places some limitations on how easily an electron can move between any given pair of states by emitting light. For certain "forbidden" transitions,* this can increase the lifetime of the higher energy state by enormous factors: some strongly forbidden transitions with frequencies in the visible range of the spectrum can have lifetimes of tens or even hundreds of seconds, making them potentially useful as clocks.

While the need for a long lifetime tends to argue for basing a clock on a pair of states with a relatively low transition frequency, once the lifetime gets up to tens of seconds, the spontaneous emission issue ceases to be much of a factor.† At this point, measurement concerns argue for using the *highest* frequency possible for the measurement. We can understand this in terms of our basic model of timekeeping as the counting of ticks: assuming we can keep an accurate count to some level—say +/− 1 tick—we can improve the precision of our measurements by increasing the number of ticks between measurements. If we can measure the frequency of light to within +/− 1 Hz, the relative uncertainty in the measurement will get smaller as the frequency gets larger. A clock based on a transition with a frequency of 1,000,000,000 Hz will have an accuracy of 1 part per billion, corresponding to about 0.03 seconds per year. Increasing the

---

* These transitions generally involve moving an electron between two states with a difference in angular momentum that can't be accounted for by the angular momentum of the emitted or absorbed photon (which usually carries one unit of angular momentum). The transition can still happen but involves more complicated processes that are much less likely, thus greatly decreasing the rate of emission.

† The roughly one-second time between Ramsey measurements in a cesium fountain clock is the longest of any clock system; most next-generation clocks use cycles of a tenth of a second or less. A lifetime of tens of seconds is essentially forever compared to this, and there's no significant risk of spontaneous decay throwing the clock off.

frequency to 10,000,000,000 Hz, all else being equal, improves the accuracy to 0.1 parts per billion, or 0.003 seconds per year.

Cesium is a good choice for an atom based on both of these criteria—the lifetime of the upper state for the clock transition is many years, and the frequency of the transition is the highest for this type of transition in any naturally occurring element. It is also a good fit for the technology of the time at which atomic clocks were first developed, benefitting enormously from the rapid advances in the technology for generating and manipulating microwaves that were driven by the radar development projects during World War II. The 9.192 GHz frequency of the cesium clock transition is relatively high, but there are straightforward electronic techniques for reducing the frequency to a point where the oscillations can be counted and used to drive a clock display.

their success in the lab, one of Maiman's assistants, Irnee D'Haenens, jokingly called the laser "a solution looking for a problem." The intervening 60 years have turned up any number of problems that can be solved with the use of lasers.

Lasers, like atomic clocks, are based on Bohr's idea of allowed states for electrons in atoms, with light being absorbed or emitted as electrons move between states. The other key insight needed came from Albert Einstein, in a groundbreaking 1917 study of the statistical properties of this absorption and emission: in addition to the well-known processes of spontaneous emission (where an atom in a high-energy state emits a photon and drops to a lower energy state) and absorption (where an atom in a low-energy state absorbs a photon and moves to a higher energy state), Einstein introduced a new process called "stimulated emission." In this process, an atom in a high-energy state encountering a photon of the correct frequency can be "stimulated" by its interaction with the light to emit a new photon that exactly matches the first in frequency and direction.

Stimulated emission makes it possible to amplify light of a particular frequency, a process first demonstrated by Townes in 1953 in a process he called Microwave Amplification by Stimulated Emission of Radiation. He prepared a steady stream of ammonia molecules in a high-energy state and sent the beam through a microwave cavity. When a molecule decayed inside the cavity, the emitted photon would become trapped inside, bouncing back and forth between the walls. Later high-energy molecules were more likely to encounter these trapped photons and be stimulated to emit new photons of the same frequency; over time, the number of photons inside would increase exponentially, making a source of microwaves with an exceptionally small spread in frequencies.

The "maser" (named from Townes's acronym) was a useful tool for microwave spectroscopy, and it also formed the basis for a prototype molecular clock based on the transition between two states in ammonia. While the performance of cesium clocks ultimately makes them better for defining the second, hydrogen masers using the microwaves associated with the two lowest energy states in hydrogen are a key part of the ensemble of time standards used to assemble UTC (as discussed in Chapter 13).

Not long after developing the maser, Townes began working on ways to make an "optical maser" by extending the process of amplification by stimulated emission into the range of visible light. At around the same time, Gordon Gould, then a graduate student, began considering the same process and gave Townes's acronym a small tweak, leading to the modern name: Light Amplification by Stimulated Emission of Radiation, a "laser."*

The elements of a laser are similar to the components of a maser: a gain medium containing atoms or molecules prepared in high-energy states and a cavity to catch the light they emit and direct it back through the medium to be amplified. The simplest laser configuration sandwiches a gain medium between two mirrors, reflecting photons back and forth along a line between them. The number of photons in the cavity increases exponentially: a single spontaneously emitted photon can stimulate the emission of a second photon; then the two identical photons become four, then eight, and so on. No real mirror is perfectly reflective, so a tiny fraction of the light inside will leak out; since all of the photons created by stimulated emission have exactly the same frequency and are moving in the same direction, the light emerges in the tight beam of a single color that we recognize as a laser.

The exact frequency of the light is determined by the energy difference between the high- and low-energy states involved in the stimulated emission process, which for a gas laser is fixed by the choice of element. If those atoms are embedded in a solid medium, though, interactions with other nearby atoms smear the narrow emission lines of a single atom out into broad emission "bands," spanning a range of frequencies determined by the structure of the material. A sample of a solid material induced to emit light will produce a broad spectrum covering a wide range of frequencies.

This broadening of lines into bands means that a solid gain medium can be used to make a laser that is tunable over a wide range of frequencies. The stimulated emission process will produce photons that are identical to each other, but the exact frequency amplified can be anywhere within the band. Adding other elements between the mirrors of the laser cavity allows physicists

---

* Townes was granted the initial patent for the laser, but Gould had notarized notebook pages showing that his work was actually earlier; this led to a protracted court fight about who, exactly, deserves credit for the invention. In the interest of full disclosure, I should note that Gould was a 1941 graduate of Union College, where I teach.

to select and amplify one particular frequency within that range. Such tunable lasers are the key to modern high-precision spectroscopy: rather than using a diffraction grating to passively spread out light from a natural source, as in the days of Joseph von Fraunhofer, modern spectroscopists selectively illuminate atoms with tunable lasers to identify the frequencies they are most likely to absorb. These methods can even be used to identify forbidden transitions that emit light too infrequently to be seen in a grating spectrometer: such transitions also absorb light very rarely, but a laser can produce a beam with so many photons in it that even very unlikely transitions can be excited.* As a result, laser spectroscopy has completely revolutionized the study of atoms and molecules over the last 60 years.

Lasers, Combs, and Clocks

The ability to find and excite forbidden transitions with lasers is one of the keys to making next-generation atomic clocks. In the same way that the core of a cesium clock is a highly stable microwave source, the core of an optical-frequency atomic clock is a laser whose frequency is controlled to within about 1 Hz out of around 1,000,000,000,000,000 Hz. This leaves a major problem, though, which is how to convert the incredibly high frequency of visible light, which is too high for even modern electronics to count directly, into a useful time signal. This problem, too, is solved by the use of lasers, but a very different *kind* of laser.

The above description of how a laser works implicitly assumed a continuous-wave laser, like most of the ones we encounter in laser pointers and supermarket grocery scanners. Such lasers, as the name suggests, emit a continuous beam with an exceptionally narrow spread of frequencies (compared to any non-laser source of light). The same solid media that are used to make tunable continuous lasers can be used to instead emit light in extremely short pulses with an exceptionally *broad* spread of frequencies. While this might seem an unlikely item to be of use in precision timekeeping, in fact these pulsed lasers are essential for getting a low-frequency time signal from a visible-light clock.

---

* Once excited to the high-energy state via the forbidden transition, these atoms can be induced to return to the lowest energy state by way of a transition to some intermediate state. This involves emitting a photon of a very different color than the ones from the laser, so spectroscopists look for those as a sign that the absorption has happened.

It's easiest to understand the operation of a pulsed laser by considering a cavity in a ring configuration, where light passes around in a loop and always enters the gain medium in the same direction. As we know from relativity, the speed of light is constant and finite, so there is a particular time required for the light to complete one loop and return to the gain medium. This is determined by the length of the path the light follows through the cavity; it also sets the repetition rate of the laser (since the pulse traveling around the cavity will strike the output mirror once per loop).

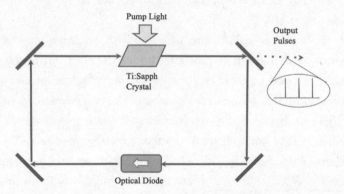

Schematic of a ring-laser cavity. Light at a broad range of frequencies is emitted and amplified by atoms of titanium in a sapphire crystal excited by light from a pump laser. An optical diode allows light to circulate in the cavity only in one direction, and interference between the different frequencies produces a train of very short pulses of light; some of this light leaks out through one of the cavity mirrors.

This time plays a critical role in determining the frequency of the output light because of the wave interference phenomenon we discussed when talking about the Michelson Interferometer back in Chapter 12. If the time needed to come back to the gain medium is an integer multiple of the oscillation frequency, when the light comes back around, the peaks of the returning wave will fall on top of the peaks of the newly emitted wave, adding together constructively to make a more intense light at that frequency. If the frequency is a little bit off, though, waves that have made different numbers of trips around the loop will arrive with different phases and destructively interfere, canceling each other out. The laser will thus operate efficiently only at a limited number of frequencies, determined by the size of the cavity. Each of these cavity modes has a frequency that's an integer multiple of the repetition frequency of the laser.

The same basic physics comes into play even in a continuous-wave laser, where the range of frequencies used is restricted in some way, picking out a single frequency matching the characteristic frequency of the cavity. This is why the light emitted by continuous lasers is even narrower in frequency than the light emitted by a gas of atoms. If the gain medium used is one of the solid materials that can amplify an extremely wide range of frequencies, though, and nothing is done to restrict that range, the laser can operate in *all* those allowed frequencies *at the same time*.

Those different light frequencies also interfere with one another, but the effect of that is not to change the overall intensity of the laser but to make it depend on time. If you add together two waves with very slightly different frequencies, the result looks like a wave at the average of the two frequencies, modulated at the difference between the two frequencies. With sound waves, this gives the combined wave a sort of pulsating quality, getting louder and softer in a regular pattern. Musicians use this when trying to tune two instruments to the same pitch: they listen for this "beat note" between the two and try to eliminate it by adjusting the frequency of one instrument to make the beat note as slow as possible.

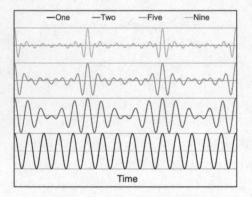

When adding more than two frequencies, the same basic thing happens, but the areas where the waves cancel out expand. In terms of sound waves, the periods of silence get longer while the audible sound gets louder, making the beating more pronounced and discordant.* In the case of light, the result of adding together

Adding waves with slightly different frequencies leads to shorter pulses as the number of frequencies involved increases from one to two, to five, and then to nine.

---

* This is why the sound of two slightly out-of-tune instruments only bothers musicians, while an entire elementary school orchestra playing slightly different notes is nearly unbearable to anybody who isn't a parent.

many different frequencies is to make pulses of a very short duration—femtoseconds or even attoseconds (one attosecond is 10^{-18} s, or 0.000000000000000001 s) for a state-of-the-art pulsed laser.

Physicists interested in the properties of atoms and molecules make use of the short duration of these pulses to study the motion of electrons on the atomic scale (an attosecond is roughly the time an electron would take to complete an orbit in a Bohr atom, so it sets the scale for atomic and molecular processes). These are also used to generate short, intense X-ray pulses that are potentially useful for studying materials and biological systems, so there is a thriving community of researchers working at the ultrafast-laser frontier.

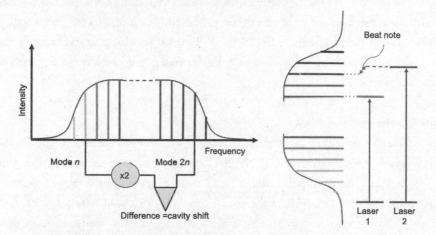

A frequency comb. Left: An octave-spanning comb allows the measurement of the cavity shift by doubling the frequency of the nth mode and comparing it to the frequency of the 2nth mode. This completely determines the frequency of any mode in the comb. Right: The frequency difference between two modes is determined by adjusting the comb so one mode exactly matches the frequency of one laser, then measuring the beat note between the other laser and another mode of the comb.

For physicists interested in the precise measurement of time, the attractive feature of these lasers is their frequency composition. A laser making pulses a few femtoseconds in length will span a full octave of frequency—the highest frequency emitted is at least double the lowest—with a "comb" of modes spaced by the repetition frequency of the laser. Since that repetition frequency is controlled by the length of the cavity, it can be made extremely stable and reliable, allowing physicists to know the frequency of any "tooth" of the comb absolutely.*

Femtosecond-pulse laser frequency combs are an invaluable tool for accomplishing precision measurements of time and frequency. They provide an easy way to connect the frequency of a laser operating in the visible range of the spectrum (whose frequency is far too fast to count directly with the electronic systems used in microwave clocks) to the repetition rate of the laser pulses (whose frequency is generally in the radio or microwave regime and thus well suited to electronic measurement). Frequency combs can also be used to directly compare the frequencies of two different visible lasers. This is done by adjusting the comb so one tooth exactly matches the frequency of one of the lasers and then comparing the frequency of the other to the tooth closest to it. This method works to compare frequencies to 17 or 18 digits of precision, a level far beyond cesium clocks, making it possible to consider redefining the second in terms of visible light.

Ion Clocks and Relativity

In addition to increasing the frequency of the light used in the clock, next-generation atomic clocks must also deal with many of the same issues that affect cesium clocks. In particular, a good atomic clock should allow for a long interrogation time to compare the atom's internal clock to the frequency source in the lab (since the accuracy achievable with the Ramsey method increases as the time between interactions gets longer) and should minimize the random motion of the atoms (which causes Doppler shifts of the frequency they absorb and emit). Both of these argue for confining the atoms to be used as the frequency standard to a very small region of space: if the atoms are held in place very tightly, that both minimizes their motion and allows them to spend a long time interacting with the radiation.

---

* This is complicated slightly by a technical detail regarding real cavities, which shifts the entire comb by some amount, so each "tooth" has a frequency that is an integer multiple of the repetition frequency plus some cavity shift. Given a wide enough comb, though, this shift can be measured directly by using a special material to produce light at double the frequency of a low-frequency mode and comparing that to the frequency of a mode at twice the frequency; the difference between the two gives the cavity shift.

Electrically neutral atoms like the cesium used in current atomic clocks, though, do not generally interact all that strongly with the outside world, making it tricky to confine them tightly. For this reason, some of the best candidates for atomic clocks that go beyond cesium are based on ions, atoms with at least one electron removed. Ions continue to have the allowed-energy-level structure that is the core principle of an atomic clock (though the allowed electron orbits in an ion have different energies than in a neutral atom of the same element), but because they are electrically charged, they interact very strongly with other electrically charged objects. This means they can be confined to a small region of space by surrounding them with high-voltage electrodes arranged to always push the ion back toward the center, forming an ion trap.* Once trapped, laser light can be used to cool the ions to the point where they are essentially motionless at the center of the trap, an ideal starting point for making an atomic clock.†

One of the best prototype clocks in the world is based on a single trapped ion of aluminum-27, built at the National Institute of Standards and Technology by the group of David Wineland.‡ The clock transition in aluminum has a frequency of 1.21 x 10^{15} Hz, a hundred thousand times larger than the cesium clock frequency, and the upper state has a lifetime of around two minutes, both excellently suited for use as an optical clock. There is one significant technical problem, though: the laser frequencies needed both for driving the clock transition and the laser cooling of the aluminum ion are in an extremely inconvenient region of the spectrum for modern laser technology, making direct cooling and measurement of the ion's state difficult. The researchers had to put in extreme effort to make a laser system that could excite the clock transition; making a second system to directly cool the aluminum would be too much.

---

* It's actually impossible to do this with a set of constant voltages, so real ion traps use high-voltage electrodes that switch from positive to negative at high frequency. These alternately pull and push on the ion so rapidly that the ion can't move very far in response, and ends up feeling an average force that confines it to a small region near the center of the trap.
† This is the same laser-cooling process used in cesium clocks. As we said in Chapter 13, it's fascinating but would require a whole book to explain in detail.
‡ Wineland shared the 2012 Nobel Prize in Physics for his work with trapped ions (the citation was "for groundbreaking experimental methods that enable measuring and manipulation of individual quantum systems" and the prize was shared with Serge Haroche, who worked on the physics of single photons trapped in cavities).

The NIST team's ingenious solution was to add a second ion (of magnesium-25) to the same trap. The two ions interact with each other in ways that depend on their relative motion, allowing the researchers to use lasers that are much easier to work with to cool the magnesium ion, and thus indirectly cool the aluminum ion through its interaction with the magnesium. The cold magnesium ion can heat back up slightly by absorbing energy from the aluminum ion, which cools down in the process. Repeating the cooling process removes this energy from the magnesium ion, which then takes some more energy from the aluminum ion, and this cycle can be repeated until both ions are equally cold.

They can also use the interaction between ions to "map" the state of the aluminum ion onto the magnesium ion, allowing them to read out the final state at the end of the atomic clock sequence. As of 2019, this "quantum logic clock" technique had resulted in an accuracy better than one part in 10^{18}; if it could be operated continuously, such a clock would need to run for 240 times the current age of the universe before gaining or losing a single second!

Maybe the most compelling demonstration of the power of these clocks was an experiment done just over a decade ago, in 2010, with an earlier version of the aluminum clocks. The NIST team built two nearly identical clock systems and compared them under different conditions to verify the predictions of relativity at unprecedented precision.

In the first experiment, they applied an extra voltage to the ion trap in one of the two clocks to shift the ion slightly away from the center of the trap. This caused it to oscillate back and forth in the trap, with an average speed they could vary by changing the applied voltage. (The moving ion acts like a moving clock, and as we saw in Chapter 12, moving clocks run slow.) This allowed them to directly test the predictions of special relativity by comparing the number of ticks in one second on the stationary clock to the number of ticks for the moving ion clock.

The velocities involved here are very slow, down to just a few meters per second, an easy walking pace, but the clocks were sensitive enough to detect the effect of relativity. At the highest speed they measured, a bit over 35 m/s (on the high side of highway driving speed), the shift amounted to just 8 ticks out of 1,210,000,000,000,000, but is unmistakable in the data. The change in the moving ion's tick exactly matches the prediction of special relativity, showing that Einstein's theory holds even for motion at everyday speeds.

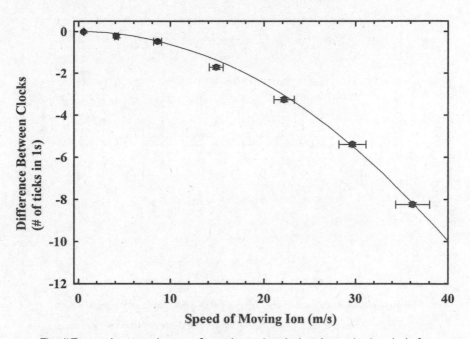

The difference between the rate of a stationary ion clock and a moving ion clock, for different speeds. The solid line shows the time dilation predicted by special relativity.*

They could also use these ultra-precise clocks to test general relativity's predictions regarding time and gravity. For this second experiment, they set up the two clocks in neighboring labs and made repeated measurements of the frequency difference between them over a period of several days. Then they used hydraulic jacks to lift up the table holding one of the clocks by about 33 centimeters (just over a foot) and made several more measurements of the frequency difference with one clock at higher altitude than the other.

In this case, general relativity (discussed in Chapter 14) predicts that the elevated clock should tick slightly faster, and again, the result is clear in the data. The shift here is even smaller—just five ticks out of 121,000,000,000,000,000—but clearly detectable. And, again, the measured shift agrees beautifully with the prediction from Einstein's theory.

---

* Illustrations on pages 303 and 304 modified from C. Chou, Hume, D., Rosenband, T., and Wineland, D., "Optical Clocks and Relativity," *Science*, 329 (5999), 1630–1633, (2010), DOI: 10.1126/science.1192720. Used with permission.

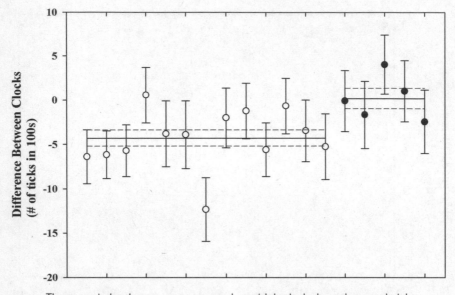

The open circles show measurements taken with both clocks at the same height;
the filled circles indicate measurements with one clock raised 33 centimeters
above the other. The horizontal lines represent the average of the different
data sets, with the dashed lines showing the uncertainty in the averages.

These experiments from 2010 used an earlier model of the aluminum ion clock, which has improved substantially since then, though they have not repeated the test of gravity with the latest model. The other prototype clock technology currently being developed, though, has been used to measure the gravitational shift at even higher precision, using a landmark feature of the Tokyo skyline.

JEAN RICHER IN THE TWENTY-FIRST CENTURY: OPTICAL LATTICE CLOCKS

As noted above, the great advantage of using ions is that their electrical charge makes it easy to push them around with high-voltage electrodes and trap them very tightly. Neutral atoms are harder to exert forces on, but it's still possible to trap them. In particular, atoms can be trapped in the focus of an intense laser beam tuned to a frequency slightly lower than that needed to excite the atoms to a higher energy state. This "optical tweezer" allows single atoms (and larger but still-microscopic objects) to be moved around with forces exerted by light; Arthur Ashkin won a share of the 2018 Nobel Prize for developing this technique.

The force that pulls atoms into the center of a bright beam can also be used to trap large numbers of atoms in an interesting way. If instead of using a single focused laser, we use two lasers with the same frequency traveling in opposite directions, the two waves will interfere with each other, making a standing wave pattern that remains fixed in position. There will be places where the waves destructively interfere, making dark spots where there is no light, and places where the waves constructively interfere, making bright spots. Each of these bright spots can serve as a tiny trap for a single atom, with the individual traps separated by one-half the wavelength of the light.

If we add more lasers, we can turn this into a three-dimensional array of tiny atom traps.* The trapping sites in such an "optical lattice" behave like atoms in a crystal, separated by about half the laser wavelength, with the cold atoms added to the lattice acting like electrons in a solid.†

Atoms trapped in these lattices can be confined and laser cooled to extremely low temperatures, much like trapped ions, which at first glance may seem like a good candidate for an atomic clock. On closer investigation, though, there is a technical issue: the force that traps the atoms comes from a "light shift" that reduces the energy of the ground state. This energy shift also changes the transition frequencies of all the other states, so while the atoms are held in the lattice, the frequency of the light needed to move between the clock states is different than it would be for atoms "in the dark." The clock frequency would then depend on the exact frequency and intensity of the lattice light, and a clock whose ticking frequency depends on the details of the lattice is not a standard that can compete with cesium fountain clocks.

---

* The most obvious way to do this is to use three pairs of beams along the three directions of space (north-south, east-west, and up-down), trapping atoms where the bright spots from all three interference patterns coincide. This can also be done with just four laser beams, with the right choice of angles.

† In fact, there is a very active field of research dedicated to investigating this analogy, using the motion of atoms in an optical lattice to explore the physics of electrons in a solid; this has connections to phenomena like superconductivity, which are fascinating but not directly relevant to this book.

　　If we look even closer at the details of the interaction between the atoms and the light, though, we can find a path back to the "at first glance" situation. The light shift of a particular energy state depends on all the possible transitions from that state to any other, and the shifts from different transitions can act in different directions. These shifts can be played off against one another, and given just the right combination of states and choice of laser wavelength, they can cancel out. There can be a "magic wavelength" in which the lattice light shifts the energy of the upper state of the clock transition by exactly the same amount as the shift in the energy of the ground state that creates the lattice in the first place.

　　Atoms trapped in a magic wavelength lattice are thus an ideal basis for an atomic clock: they are tightly confined by the lattice laser, allowing long interaction times and minimizing Doppler shifts caused by the motion of the atoms. From the perspective of the clock transition, though, they behave as if they were in the dark, giving a reliable frequency reference that can be used to drive a clock. The magic wavelength idea was first developed by Hidetoshi Katori of the University of Tokyo, in proposing an atomic clock using strontium atoms in an optical lattice. The clock transition in strontium corresponds to a wavelength of about 698 nanometers (a deep red, just barely in the visible spectrum), while the magic wavelength for the optical lattice is 813 nanometers (just outside the visible range in the near-infrared). The Katori group has been operating strontium lattice clocks since around 2005, and numerous other groups around the world have joined them.

　　Clocks based on atoms in optical lattices offer one significant advantage over those based on trapped ions, namely that the lattice clocks involve samples of several thousand atoms. This allows the transition probability to be determined more quickly and accurately than for a single-ion clock, where the probability must be determined by repeating the clock cycle multiple times. As a result, lattice clocks have a slight performance edge, and over the last several years, the title of "world's most accurate clock" has been traded between lattice clocks in different standards laboratories.

　　As a dramatic demonstration of the power of these clocks, in 2019 the Katori group took two of their optical lattice clocks out of their lab and set them up at the Tokyo Skytree tower. They kept one clock at ground level and put the other on the observation deck, some 450 meters above. Over the course of a week, they made 11 measurements comparing the frequency of the two clock lasers and found that they differed by about 21.177 Hz, out of a laser frequency of 429,228,004,229,952 Hz. This matches the time dilation predicted from general relativity for the height of the tower, measured to a precision of a few centimeters.*

The measurements in both Boulder and Tokyo are spiritual heirs to the seconds pendulum measurement of Jean Richer, back in the 1670s (described in Chapter 6). Richer's measurements were a key milestone in the field of geodesy, showing a slight weakening of the force of gravity due to the equatorial bulge of the earth, which increases its radius at the equator by about 10 kilometers relative to Richer's home base in Paris. The optical-clock measurements described here use a much subtler gravitational effect, a change in the structure of space-time itself. Thanks to the phenomenal precision of atomic clocks, though, they can detect a difference in elevation of less than 10 *centimeters*. This opens some exciting potential for measuring the shape of the earth on a very fine scale, and for monitoring it in (pardon the pun) real time as it changes due to catastrophic events such as earthquakes—or slower ones like the movement of magma underground.

The clock-based geodesy measurements described above are just one of the applications physicists have for ultra-precise timekeepers, and a relatively grounded one at that. Comparisons between atomic clocks based on different elements can be used to check whether the fundamental constants of nature—numbers like the charge of an electron, Planck's constant, or the speed of light—are truly constant, or if they change slowly over time as predicted in some exotic theories of physics.† Our best current measurements of the possible

---

* They used both laser ranging and satellite measurements to monitor the height of the tower during the demonstrations, which was constant to within a few centimeters, as expected given the expansion and contraction of the tower itself due to changes in the weather.
† For technical reasons, what really matters are ratios of these constants to each other, such as the "fine structure constant," which describes the strength of electromagnetic interactions and thus plays a key role in determining the energies of electron states in atoms. This divides the Coulomb constant for the strength of electric forces multiplied by the electron charge squared by the product of Planck's constant and the speed of light. Different researchers searching for changes in the fine structure constant will offer various explanations as to which of the underlying numbers is "really" changing—what one would call a decrease in charge, another would call an increase in the speed of light—ensuring that even if a change is someday definitively detected, physicists will still have something to argue about.

drift in these constants mostly rely on looking backward in time—for example, studying light emitted billions of years ago by atoms in distant galaxies to see if the spectra match those of modern-day atoms of the same element. Changes in the constants would shift different atomic states by different amounts, though, so comparing the frequencies of two modern-day clocks based on different atoms can tell us whether the constants are changing *right now*. A change of 10 parts per billion over the course of a billion years (comparable to the best historical measurements) is equivalent to an average change of a few parts in 10^{17} per year; that same level of sensitivity was reached by comparing the frequency of the aluminum-ion clock described above to that of a clock based on an ion of mercury for just one year, back in 2007. The best current measurement comes from a comparison of two different "clock" transitions in an ion of ytterbium and is better by about a factor of 10. They haven't seen any changes yet, but as clocks get better and the experiments run for longer, the sensitivity keeps increasing, so if the constants of nature aren't *really* constant, atomic clocks will let us know.

Another possibility that has been floated involves using a whole network of precision atomic clocks to detect more fleeting shifts in the clock states due to gravitational waves. A passing gravitational wave causes a stretching and compression of space; these have been detected by the Laser Interferometer Gravitational-Wave Observatory (LIGO) using a Michelson interferometer whose arms are several kilometers in length. These waves should also cause small shifts in the rate at which clocks tick, getting slightly faster and then slightly slower (or vice versa) as the wave passes. A large number of ultra-precise atomic clocks in different locations, either on the earth or on satellites in space, constantly monitoring each clock's frequency relative to the others, could watch the passing wave as a time shift rippling through the network. The LIGO detectors excel at finding spectacular collisions between black holes; a clock network would be sensitive to more subtle events that could tell us about the conditions of the universe in the instants immediately after the Big Bang.

These kinds of clock comparisons have also been proposed as a way to detect the invisible "dark matter" believed to account for most of the mass of the universe. Some of the candidates for dark matter particles would interact with atoms to create a slowly oscillating change in the energies of atomic states. This would depend on the details of the particular atoms and states involved, and thus it could be picked up by comparing clocks based on different atoms. If an aluminum-ion clock ticks first slower and then faster than a strontium-lattice clock, that might be the result of dark matter. Other models of dark matter could cause more fleeting shifts as clumps of dark matter pass near the earth, which might be detected as a change spread across a network of clocks, much as with gravitational waves.

TIME FOR THE FUTURE

Of course, none of these exotic physics effects are going to impair your ability to use GPS navigation, let alone make you late for an afternoon meeting or dinner with friends. The shifts these would cause are still beyond the sensitivity of the world's best experimental atomic clocks, let alone the workhorse standards that define time for the world.

This is very much in keeping with the long tradition of timekeeping development, though. Scientists and engineers are never happy with the best that they have and are always pushing forward, imagining the amazing things that could be done if they had a clock that was just a *little* bit better. We saw this directly with Galileo Galilei proposing using Jupiter's moons as a clock 40 years before Jean-Dominique Cassini and Ole Rømer made it a reality, or Gemma Frisius proposing a chronometer to keep track of longitude a full 200 years before John Harrison. It's even visible in the tomb inscriptions of the Egyptian courtier Amenemhet in 1500 BCE, touting his invention of a water clock that measures the unequal hours of the night in different seasons.

That push for ever-greater precision has been a spur to innovation for thousands of years, both for individual inventors whose fame is well recorded like Christiaan Huygens and Su Song, and for the anonymous tinkerers around 1300 CE who invented the first sandglass or the first mechanical clock. Their names may be lost to history, but their work moved science and technology forward and provided a solid foundation for later discoveries.

This drive toward better measurements is also reflected in more fundamental science, in the need to impose order on the complicated patterns of the natural world so as to better forecast the future. Whether it's Simon Newcomb compiling the *Tables of the Sun*, Tycho Brahe and Johannes Kepler plotting the orbit of Mars, the Mayan astronomer-priests tracking the rising of Venus, or even the builders of Newgrange piling up tons of stone to mark the winter solstice, astronomers have spent millennia refining their observations in order to know not just what time it is now, but what time it *will be* in the future when astronomical cycles repeat themselves.

So, while existing clocks are more than sufficient for the timing of everyday events and navigation using GPS, it is no surprise that physicists continue to push the frontier of time measurement. The millennia-long tradition of humans designing and building ever-better clocks shows no sign of coming to an end anytime soon.

Acknowledgments

As mentioned in the text, this book is based in part on a course I have taught on the history of timekeeping. I'm very grateful to my colleagues at Union College for the opportunity, in particular to the faculty of the Department of Physics and Astronomy, to Maggie Graham and the Scholars program, and to Anastasia Pease for asking me to do a guest lecture in a different course that provided the inspiration for the whole thing.

I'm also grateful to more distant colleagues who have helped me become better informed about various aspects of the subject, or at least helped me avoid saying things that are actively dumb. In particular, Thony Christie of "The Renaissance Mathematicus" and Tom Swanson of the US Naval Observatory provided invaluable assistance. Anything I got wrong about either Galileo or atomic clocks is entirely my fault, despite their very best efforts.

Getting from the initial idea to final book involves the work of many hands, and what you're reading is substantially better thanks to their efforts. First and foremost, thanks to Alexa Stevenson and Laurel Leigh for editing this book, and to my agent Erin Hosier for help refining the original idea. Thanks, too, to Sarah Avinger and Jessika Rieck for making the cover and interior layout look great, and Scott Calamar for the copyedits that make me look like someone who understands and can consistently apply the rules of English grammar. And everyone else at BenBella Books, who have been a real pleasure to work with.

Almost every author who puts fingers to keyboard to write an "Acknowledgments" section for a new book talks about what a challenge it was to write the book in question; I've done it myself. This one might be the most extreme possible version of this, as I was about four chapters into the first draft in early

March of 2020 when the COVID-19 pandemic abruptly shut everything down. I am incredibly fortunate in that everything I needed to do for both my job as a professor and the writing of this book could be handled remotely, but even so this was an incredibly difficult and stressful time, a set of experiences that I hope never to repeat. A long list of people deserve at least some credit for helping me make it through this process with my sanity (reasonably) intact: the Fugitive Popes on Zoom, the Non-Student Happy Hour crew at Union (both remotely and in person), the staff at Hunter's on Jay, and the folks at the Starbucks in Niskayuna.

Most of all, though, this would've been absolutely impossible without the love and support of my family, particularly my parents, Ron and Jan Orzel; my kids, Claire and David; and most importantly my wife, Kate Nepveu. I love you all, and couldn't have done it without you.

Recommended Reading

If this book has piqued your interest in the topic of timekeeping, or any of the subtopics covered within these chapters, there are an oversized handful of books that were especially useful in the course of my research that I would recommend to anyone interested in learning more. Four of these offer broad historical surveys; in alphabetical order by author, these are:

Empires of Time: Calendars, Clocks, and Cultures by Anthony Aveni (Basic Books, 1989)

Calendar: Humanity's Epic Struggle to Determine a True and Accurate Year by David Ewing Duncan (Avon, 1988)

In Search of Time: The History, Physics, and Philosophy of Time by Dan Falk (Thomas Dunne, 2008)

Revolution in Time: Clocks and the Making of the Modern World by David Landes (Belknap Press, 1983)

I also recommend three books with a narrower topical focus:

The Quest for Longitude (proceedings of the Longitude Symposium of 1993) edited by William J. H. Andrewes (Harvard University, 1993)

Tycho & Kepler: The Unlikely Partnership That Forever Changed Our Understanding of the Heavens by Kitty Ferguson (Walker and Company, 2002)

Einstein's Clocks, Poincaré's Maps: Empires of Time by Peter Galison (Norton, 2003)

As I hope you've seen from reading this book, the history of timekeeping is an enormous topic, spanning all of recorded history and a bit beyond, so this list is by no means complete. However, these books (and the references therein) provide an excellent place to start a deeper exploration of the subject.

Index

About the Author

Photo by Matt Milless

Chad Orzel is a professor at Union College in Schenectady, NY, and the author of four previous books explaining science for nonscientists: *How to Teach Quantum Physics to Your Dog* (Scribner, 2009) and *How to Teach Relativity to Your Dog* (Basic Books, 2012), which explain modern physics through imaginary conversations with Emmy, his German shepherd; *Eureka: Discovering Your Inner Scientist* (Basic Books, 2014), on the role of scientific thinking in everyday life; and *Breakfast with Einstein: The Exotic Physics of Everyday Objects* (BenBella, 2018), on the ways phenomena from quantum mechanics manifest in the course of ordinary morning activities. He has a BA in Physics from Williams College and a PhD in Chemical Physics from the University of Maryland, College Park, where he did his thesis research on collisions of laser-cooled atoms at the National Institute of Standards and Technology in the lab of Bill Phillips, who shared the 1997 Nobel Prize in physics (not for anything Chad did, but it was a fun time to be in that group). He has been blogging about science and other things since 2002 (on his own site, at scienceblogs.com, and most recently for *Forbes* and on Substack), and occasionally appears as a talking head on TV shows about science. He lives in Niskayuna, NY, with his wife Kate Nepveu, their two children, and their new dog Charlie the pupper.